THE MENSA GENIUS QUIZ BOOK 2

The MENSA Genius Quiz Book 2

Marvin Grosswirth,

Dr. Abbie Salny, and

the Members of Mensa

ADDISON-WESLEY PUBLISHING COMPANY

Reading, Massachusetts   Menlo Park, California   New York
Don Mills, Ontario   Wokingham, England   Amsterdam   Bonn
Sidney   Singapore   Tokyo   Madrid   San Juan   Paris
Seoul   Milan   Mexico City   Taipei

Library of Congress Cataloging in Publication Data

Grosswirth, Marvin, 1931–
   The Mensa genius quiz book 2.

   1. Puzzles.   2. Questions and answers.   I. Salny,
Abbie F.   II. Mensa.   III. Title.
GV1493.G762   1983   031'.02   83-7244
ISBN 0-201-05958-4 (pbk.)

Set in 11-point Century Schoolbook by TriStar Graphics,
Minneapolis, MN
Cover design by Marshall Henrichs
Interior art by Nancy Goryl

20 21 22 23 24 25 - DO - 95949392
Twentieth printing, September 1992

Dedicated to
Jerry Salny
and
Marilyn Grosswirth
for all the obvious reasons

Contents

Acknowledgments

To enable the reader to compare his or her results with those of a group of Mensa members, those quizzes which are to be scored were given to approximately a hundred Mensa volunteers at various meetings and gatherings in the United States; London, England; and Toronto, Canada.

Among the test takers were Sharon Bailly, Judy Bank, Steve Bank, Judy Benevy, Carol Bohlen, Mickey Bregman, Vivian Bregman, Felix Bremy, Betty Claire, Seth Cohen, Mitchell Darer, Walter D'Ull, Paul Emmons, Bob Finnegan, Harold Fleming, Tracy Franz, Mary Haigh, Peter Heinlein, Sue Kent, Keith Kizer, Anne Koval, William Krause, Landon McDonald, Paul McGuffin, Melinda Maidens, Leo Marazzo, Elizabeth Marine, Isaac Marine, Allen Neuner, Jim Parsons, George Reisman, Wendy Sailer, Dan Schechter, Is Weiner, Martha Young, Carl Zipperle, all of American Mensa, and . . .

Helge Totland, Norway; Rhys Baron, John Darlow, Ian McLaren, Marion Maté, Susan Watkin, Brian Yare, British Isles; Elizabeth Hickman and Chris Spiegel, Australia; Barbara Thomson, New Zealand; J. A. Barker, Graham T. Coy, Douglas Skrelky, Sue Sparrow, Canada.

Even a quick glance will reveal that there are fewer than a hundred names here. As good as we claim to be at puzzles, we were unable to decipher some of the signatures (which may account for some misspellings in the list above, for which we apologize), especially those that were missing.

Identified or not, we are grateful to all who participated.

A number of Mensans offered questions for inclusion in the book. Our special thanks to all of them, including Mike Tuchman, Warren Spears, and R. B. Lehr, and to Rush Washburne, puzzle editor of the *Mensa Bulletin,* for his special help.

Once the questions were all in, Jerome E. Salny (who claims he enjoys quizzes and was in no way coerced by the quizmaker who lives with him) proofread all of them the hard way—he worked through them.

Throughout the book, in the sections between the warm-ups and the actual tests, you'll find examples of how Mensans use their brains in practical ways. These are all actual situations, contributed by members (all of whom are identified). These are only a small portion of the wealth of stories we received. Unfortunately, space considerations do not allow us to publish them all, but we are grateful to everyone who contributed (and we are saving their submissions, should there ever be a *Mensa Genius Quiz Book 3*).

We received those stories because most of the more than a hundred local-group newsletter editors published our appeal for anecdotes. Without that sort of cooperation, this book would not have been possible. The distribution of that appeal was accomplished through the small but efficient staff at Mensa's National Office, headed by Executive Director Margot Seitelman. The American Mensa Committee (our Board of Directors), and its chairman, Gabriel Werba, gave the support, encouragement, and cooperation we needed to proceed with and complete this project.

We would be remiss in not thanking the readers of the first *Mensa Genius Quiz Book*. We're beginning to realize that gremlins are no myth. Despite our, and our editors', efforts to achieve perfection, those rotten little creatures sneak into the press room just before printing begins and move things around (which is why—if you believe the answer to one of the questions in that book—you now believe that Lassie is a whale). Of course, we received letters from our readers. Every one of them was a joy, because they were all written in the good-natured spirit of fun that we hoped would prevail in that book— and in this one.

The Mensa Genius Quiz Book 2 really is the work of the members of Mensa. All we've done is pull it together. Nevertheless, in that pulling together, we recognize that any errors and omissions that may have crept in are our full responsibility. (For the record, Salny is the question maven; Grosswirth worked on the text sections.) We cheerfully accept that responsibility, gremlins notwithstanding.

All we ask is that you be gentle with us.

As with the first volume, a portion of the proceeds of this book go to the Mensa Scholarship Fund.

MARVIN GROSSWIRTH

February 1983   ABBIE F. SALNY

THE MENSA GENIUS
QUIZ BOOK 2

Are You a Genius?

Have we met somewhere before—perhaps in the pages of the first *Mensa Genius Quiz Book*? If so, welcome back. And feel free to skip most of this introduction.

If, however, this is our first encounter, stay with us for the next few pages. The primary purpose of this little book is to provide pleasure and diversion for people who enjoy puzzles and quizzes, but it can also offer some hidden surprises and rewards. You may discover that you're a potential genius.

A GENIUS—YOU?

It's entirely possible. First, let's clear up some of the doubts many people have about the word, beginning with the canard that "genius borders on insanity." High intelligence is no more of an option on a ticket to the booby hatch than is profound stupidity, or any

1

other mental extreme. Genius isn't so much a question of how smart a person may be as measured by a score on some test, as it is a question of how that smartness manifests itself. Some of the dearest, kindest, sweetest people we know can't read a postage stamp without moving their lips, and some of the most despicable villains in the world have soaring IQs.

Professions, occupations, and hobbies are not necessarily indicators of intelligence either. The only requirement for Mensa membership is a score on a recognized IQ test at or above the 98th percentile—in other words, in the top 2 percent of the population. (That works out to a minimum score of 130 on the most current Wechsler scales and to 132 on the Stanford-Binet, the two major sets of IQ tests.) But that membership is a varied and eclectic one. Authors Isaac Asimov and Leslie Charteris (author of "The Saint" mysteries) are members. Carolina Varga Dinicu, who, as "Morocco," performs Middle Eastern dances, is a member. So are Adam Osborne (who designed the Osborne portable computer) and Clive Sinclair (developer of the Timex/Sinclair computer). The president of the Ford Motor Company is a member, as is the president of a six-person janitorial service in New England. Members include homemakers, computer programmers, shop owners, shop clerks, postal employees, telephone linespeople, singers, physicists, journalists, doctors, lawyers, beggars, thieves, and, for all we know, an Indian chief (although, admittedly, at this writing, none has come forth). We have soldiers and sailors, teachers and students, artists and carpenters, Republicans, Democrats, libertarians, born-again Christians, atheists, agnostics, apostates, people with ten or twelve degrees, and high school dropouts.

So how does one identify a "genius"? Well, for openers, you might try checking the several dictionary definitions of the word and see if any of them applies to you. If you have a particular talent at which you truly excel, you may be a genius at the piano, or in labor negotiations, or in concocting exotic recipes. And if your IQ is in the top 2 percent of the population, then you're certainly what psychometrists would call a "genius."

WHO CARES?

Good question. If you don't, no one else does either. Anyone who's been around Mensa long enough will eventually hear some old-timer mutter, "IQ isn't everything, but it isn't nothing either." By itself, intelligence is as useless as a fine singing voice that is exercised only in the shower. As we said earlier, it isn't how smart you are, it's what you do with your brains that matters. But first you have to know what your potential is. That's where IQ testing comes in.

Everybody knows that IQ tests have flaws and faults. Nevertheless, it's the belief in Mensa that such tests are the only tool available for measuring one's intelligence potential, and it's as worthwhile knowing what that potential is as it is knowing your height, weight, and blood type. And that's where this book can be of help.

IS THIS BOOK AN IQ TEST?

Absolutely not! IQ tests are written, normed (that is, given to groups of people to determine how to judge scoring levels), and eventually administered under carefully controlled conditions. Many of the questions in this book, however, are similar to the types of questions found on IQ tests, so if you score well on these quizzes, chances are you'll come out well on a standardized IQ test, too. Furthermore—and perhaps more important—you can compare your results with those of the hundred or so Mensans who also worked these quizzes. That could indicate whether you're a likely candidate for Mensa.

SUPPOSE YOU FAIL?

Forget it; you can't "fail." No one "fails" or "passes" an IQ test any more than anyone fails or passes a height, weight, or vision test, or any other test that measures a physical or mental attribute. Besides, this isn't a test. The only way you can "fail" is by failing to have fun with this book, in which case the failure is ours, not yours.

Furthermore—and this is probably the most serious criticism of IQ tests—many people, including some of the most brilliant geniuses of our time, are simply no good at taking tests. (There are, however, tips to test taking; see "How to Take a Test—Any Test," page 9.)

WHERE DO YOU GO FROM HERE?

If you have no trouble with tests, proceed to the quizzes themselves. They've been organized into categories that are likely to appear, with variations, on many IQ tests.

At the beginning of each section is a set of "Warm-Up" questions, designed to get you in the mood. As with any warm-up exercise, it's a good idea to take it easy and proceed in a leisurely manner at first.

In the first *Mensa Genius Quiz Book*, we separated the warm-ups from the scored quizzes with little essays about how Mensa members exercise, polish, and have fun with their intelligence. In this book, we've devoted that space to showing how Mensans apply their thinking abilities to real-life situations. We've tried to use anecdotes that relate to the particular section of questions, but in all honesty, we must admit that we're stretching a little in some cases because most of the stories are really examples of logical thinking. (And just so you can prove what you've always suspected—that there's such a thing as being too smart—don't miss the epilogue.)

When you plunge into the scored quizzes, you should time yourself carefully because here is where you'll be pitting your wits against the Mensans who took the same quizzes. There are two good reasons for keeping track of the time. If you do as well as the

Mensans—that is, if you get as many right answers within about the same time—then you'll have a good indication of how you stack up against people whose IQs have been certified by valid tests. If, however, you don't do as well—that is, if you get as many right answers, but it took you longer to get them—that could indicate that you don't work well under pressure, and that's worth knowing, too.

In any case, if you really don't care about any of this, then just work the questions, at your own pace and as the time and mood suit you, just for the sheer pleasure of it.

If you discover that you have a "genius" for coming up with alternative right answers that we may have missed, double-check your results, and if you're certain you're correct, give yourself a double score for that question. And please tell us about it.

Above all, have fun. After all, that's what this book is really all about.

This Thing Called Mensa*

In 1945, two British barristers, Roland Berrill and Dr. L. L. Ware, thought it would be an interesting experiment to gather together people of exceptionally high intelligence. Mensa was founded in London that same year, and membership was—and still is—open to anyone whose score on any one of a number of recognized standard IQ tests is in the upper 2 percent of the general population. By 1961, when American Mensa was founded, the organization had already expanded to several other countries. There are now about 70,000 members worldwide, about 48,000 of whom belong to American Mensa.

Mensa is Latin for "table," and the name was chosen because the organization was founded as and continues to be a round-table society. The symbol suggests the coming together of equals. Because of its

---

*Slightly revised from *The Mensa Genius Quiz Book* (Reading, Mass.: Addison-Wesley Publishing Company, 1981).

membership requirement, however, Mensa is often accused of elitism. In fact, it's no more "elite" than any other organization that has a requirement for membership, whether it's the American Legion, the Daughters of the American Revolution, the Actors Guild, the Authors League, or the plumbers' union. In America, there are more than 120 chapters of Mensa (called local groups), which engage in a wide variety of activities, from parties and open houses to speakers' meetings and museum trips. It is at these functions that the notion of elitism is laid to rest. As we mentioned in the introduction, Mensans come from virtually every trade, occupation, business, and profession. "If we're elitist," a national chairman once commented, "we're the most democratic elitist organization that ever existed!"

Mostly, Mensa provides an opportunity for intelligent people to meet and exchange ideas, opinions, prejudices, fears, jokes, or recipes in an atmosphere of unrestrained mindwork. The organization has a gifted children's program, a scholarship program, and the Mensa Educational and Research Foundation. There's a national magazine, local-group newsletters, and about 200 "SIGs"—Special Interest Groups—for Mensans who want to get together, by mail or in person, to share common interests.

In a sense, it's difficult to explain Mensa. At its heart, Mensa is more of a *feeling* than anything else. It's not easy to put into words. We just know that we wouldn't want to be without it.

How to Take a Test-Any Test *

If test taking makes you a little nervous, you're probably in luck. Studies have shown that a little anxiety is actually helpful. If you go into a test situation calm and completely assured, you probably won't do as well as when you are slightly concerned. Too much anxiety, on the other hand, can slow you down and interfere with your thinking.

Try to find out in advance what sort of test you're going to be taking. If it requires factual, multiple-choice answers, the best technique is a review of facts. If the test will include essay questions, you'll need to marshal your facts, organize them into a coherent whole, and practice expressing your ideas in good, clear language and in an appropriate sequence.

Also, you should determine how the test will be scored. If it's a multiple-choice test or a true-or-false test, or a mixture of the two types, and if it is scored

*Slightly revised from *The Mensa Genius Quiz Book.*

9

on the basis of number of right answers only, then try guessing whenever you don't know the answer. Chances are your subconscious will be at work and will frequently offer a "hunch" about the correct answer. Even if it doesn't, what have you got to lose?

If, however, the test is scored so that you'll lose points for wrong answers, then first answer only those questions you're sure of. Then, if time permits, go over the ones you've skipped. Whether you answer any of the skipped ones depends largely on your gambling instincts.

If you're taking a test in which you're given a limited time in which to finish, first tackle the questions you can answer immediately, without taking time to think. After you've done that, go over the ones you skipped and answer those that require a brief moment of thought. Finally, do the ones you really have to ponder over.

Always review your work. People often find a careless error or two that can be immediately corrected. You may also find that you missed an important direction or instruction, for which you may be penalized.

And above all, remember to come to any test fresh, rested, and in as cheerful a mood as circumstances will allow. That alone can add 10 percent to your score.

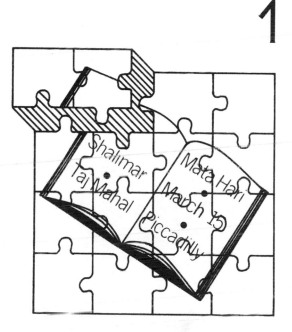

TRIVIA

Warm-Ups

1. Everyone has heard of Piccadilly in London. What does the name come from?
2. Who was the husband of the writer who wrote *Frankenstein*?
3. What is the British equivalent of West Point for training officers?
4. Sylvaner, Niagara, Sirah. What are they?
5. What is the wife of a baronet called?
6. Why can Princess Diana of England become queen some day, whereas Prince Philip can't become Queen Elizabeth's king?
7. What is a marriage between a commoner and royalty called, wherein the right of succession is abandoned?
8. The plate or badge on the bumper reads: CH. Where did the car come from? (Or at least, where did the owner buy the plate?)
9. If your baggage was labeled ORD, what airport in the United States would you expect to land at?

10. The family of a famous founder of a famous American university came from Stratford-upon-Avon. Who was the founder and what is the name of the university?

11. When the Pilgrims landed in what is now Massachusetts, they were greeted by an Indian, Squanto. What language did he speak?

12. The Plantagenets ruled England for many years. What does the name Plantagenet come from?

13. A brilliant hair color was named after an Italian painter. What is the hair color and who was the painter?

14. Where did Romeo and Juliet live?

15. She was Mrs. John Rolfe socially. Every schoolchild knows her by her first name. What is it?

16. T. E. Lawrence was strongly identified with what country?

17. According to the rhyme, what is the fate of Saturday's child?

18. Where did Peter Rabbit's father meet his doom?

19. In which of Beethoven's symphonies is Schiller's *Ode to Joy*?

20. Who was Mary Tudor's more famous sister?

21. What's the name of the drinking song whose music was used for "The Star-Spangled Banner"?

22. Where did Molly Malone of "cockles and mussels" fame live?

23. Of what country is Bernardo O'Higgins a famous national historical figure?

24. In what war did Winston Churchill serve as a war correspondent?

25. Where is the Taj Mahal?

No Small Matter

Among the various quizzes, puzzles, and contests one is likely to come across in Mensa publications and at parties, meetings, and even conventions, *trivia* questions rank high. It sometimes seems that the older a person is, the more he or she enjoys trivia quizzes, probably because trivia and nostalgia are inexorably intertwined.

It may be argued that there's not terribly much value per se in remembering the name of Gene Autry's horse (and I don't). But the simple act of attempting to conjure up that long-forgotten detail brings a flood of memories about a boyhood in which Saturday mornings were spent trudging through the neighborhood with empty milk and soda bottles in the hopes of recovering enough of the "deposit" (a kind of ransom paid to the storekeepers to ensure that the bottles would be returned) to pay the afternoon admission price to the Supreme Theatre in Brooklyn, with, perhaps, enough left over for a box of Mason's

15

Dots or Jujubes, thereby sustaining a thriving neighborhood dental practice.

However, apart from providing fun and a challenge to one's memory, and evoking frankly sentimental nostalgia, trivia has its practical aspects, too. The ability to store and recall seemingly obscure and unimportant—that is, "trivial"—facts is an integral component of a sharp intuition. If you're a veteran of the first *Mensa Genius Quiz Book,* then you are already aware of my conviction that intuition is not only a real attribute but a valuable one. (Intuition can be defined—perhaps a trifle simplistically—as the ability to rapidly "program" a set of facts randomly drawn from a vast storehouse of "trivial" data stored somewhere in the brain.)

I regard intuition as a form of intelligence. As with all forms of intelligence, people may vary in their overall intuitive capacities and skills. But, also like all forms of intelligence, intuitive abilities are not enough. To work well, even to work at all, intuition needs input, a mental storehouse full of tiny scraps of information for the mind to draw on as the need arises and circumstances dictate.

Basically, what we're talking about is a matter of magnitude. "Intelligence" requires inputs of significant data. Intuition requires inputs of seemingly insignificant data, details that you're not likely to remember consciously but that your subconscious retains and brings to the fore when a decision—major or minor—has to be made quickly.

To be sure, one person's "education" may be another person's trivia, and it isn't always easy to distinguish between the two. As an evening session student at the Baruch College of Business and Public Administration (City University of New York), I took a required course in advertising. One of the questions on the midterm examination dealt with postal regulations: "How would you mail 5,000 advertisements . . . ?" In answering, the student was expected to have cited chapter and verse of some boring postal ordinance which I could never remember. I responded with something about turning it over to the head of the mail room with instructions to let me know the fastest and most economical means of getting the mail out, pointing out that it was his job to know postal regulations, not mine. I got full credit for the answer, along with a note from the instructor requesting that I not discuss it with any of the other students.

Hans G. Frommer, American Mensa's second vice-chairman, is another person who, as a student, preferred coping with trivia to memorizing it. During his early years in engineering school, he had a chemistry professor who avoided written exams by orally testing two or three students during each period before proceeding to new material. "He would test alphabetically," Hans wrote, "so we knew about when our turn would come.

"I was an extremely lazy student," he claims, "trying to get by with a minimum of study. Consequently, I postponed studying some twenty different elements in inorganic chemistry until the last minute. Finally, I

started with the two easiest ones—gold and silver—
and got them down pat." That, he felt, would hold
him for a while. But when he showed up in class, he
made the unhappy discovery that because of several
absences, the student who was ahead of him alphabetically was being called, and Frommer would certainly
be called upon next. "I panicked," Hans admitted,
"but then I had an idea. The prof always looked at
each student's notebook and picked one of the subjects from it. I quickly creased my notebook so it
would fall open at 'Gold' and underlined the heading
with red pencil. For good measure, I did the same with
'Silver.' I had barely finished when my name was
called.

When his turn came, as he knew it must, Hans
handed his professor the notebook, "and it practically
fell open to 'Gold.' 'Well, Frommer, tell me something
about gold.' I rattled off all the data, uses, compounds,
etc., and was ready to sit down. 'Not so fast,' he said.
'That was very good, but not enough.' Again he leafed
through my notebook. It did its job and stopped at
'Silver' . . . [and] shortly thereafter I sat down with an
A and spared myself studying up on eighteen other
elements. Was it smart? I doubt it," he concluded
with appropriate self-effacement.

Perhaps the smartest thing he did was not to have
used a ring-binder type of notebook.

Some people can't seem to ignore a challenge, no
matter how trivial it may be. Consider Sandra
Stright, of Walnut Creek, California, who could have
waited a few moments to solve a minor problem but

who couldn't resist the opportunity of finding a clever way to work it out herself.

"One day at the office," she said, "I was trying to put a memo I'd written up on a metal bulletin board with a magnet. The bulletin board is too high for me to reach, so I usually get one of the taller people in the office to put things up for me, but this time no one was around who could reach it any better than I could. I could hold the paper up high enough by grasping it at its bottom edge, but I couldn't reach high enough to place the magnet."

There were three possible ways to deal with this problem: (1) Stand on something. (2) Wait for a tall person to appear. (3) Find a way that nobody else might have thought of. Clearly, the first two choices were unacceptable. After a couple of minutes' thought, inspiration struck. Instead of holding the paper to the bulletin board and then placing the magnet against the paper, Sandra attached the magnet to the staple holding the two sheets of paper together at the top, grasped the pages at the bottom, and, carefully stretching and balancing, got the magnet to touch the bulletin board. A co-worker who had observed the whole exercise characterized it as "Outrageous!" "It worked, didn't it?" was Sandra's unassailable response.

Cutting costs in a business enterprise is hardly trivial, but it sometimes takes imagination and intelligence to recognize that solutions may lie in seemingly unimportant, routine operations. As a new partner in a small typesetting company, Roberta L. Sniegocki, of

Chicago, was seeking ways of reducing expenses. One day, while reviewing recording procedures with the bookkeeper, Roberta discovered that hours were being wasted posting figures. Each employee would fill out a time sheet, from which the bookkeeper would transfer the numbers for per-job costing and billing.

Roberta devised a simple method of record keeping by designing a form that "traveled" with a job through each of the shop's departments; employees made appropriate entries as the job entered and left their section. This method eliminated considerable copying, transferring, calculating, and, not incidentally, margin of error. "After some argument about what 'everyone else' in this business does, . . . I put the new system to work. And," she admits freely, "it's been great. Now the bookkeeper has time to do all the Xeroxing, make trips to the post office—even water the plants."

What does a manufacturer do with unusable scraps? A simple—even trivial—question. "Unusable" means there is no market for them. There is, therefore, only one answer: throw them away. But people like Kathy Jones tend to be dissatisfied with simple answers, especially if such solutions are aesthetically offensive.

Kathy is a principal in Kadon Enterprises, Inc., a company that manufactures Quintillions, which Kadon promotes as "the aristocrat of strategy games." The company's advertising describes the game best:

Based on a geometric phenomenon recognized for

thousands of years—the 12 possible ways to join 5 squares—QUINTILLIONS moves these famous shapes into the third dimension and adds five fascinating games . . . plus hundreds of super puzzle challenges in 2-D and 3-D.

The sets are laser-cut from high-quality, matched-grain hardwood. There are, of course, scraps. But what aesthete could bear to part with "scraps" that also happen to be beautifully cut pieces of fine wood? Certainly not Kathy Jones and company. The solution? On the same sheet describing the game is an advertisement for "Quint-Art Sculptures—certified one-of-a-kind originals, signed and numbered editions to titillate the eye and the mind . . . in solid hardwood, permanently bonded into intriguing abstracts, witty animal and people forms. Special themes or subjects available by commission."

"I have a lot of . . . trivial items of no particular interest," Kathy said, "just routine craziness, like hanging wastebaskets on the wall so the dog won't raid them. . . ."

To fully appreciate the mental and artistic dexterity with which Kathy treats "trivialities," send a stamped self-addressed envelope to Kadon at 1227 Lorene Drive, Pasadena, MD 21122, for a copy of the company's flyer.

And if you now fully appreciate the uses of trivia, proceed to the quiz on the next page.

Match Wits with Mensa

TRIVIA TEST

Time started _____
Time elapsed _____

1. Where are the Islets of Langerhans?
2. Margaretha Geertruida Zelle (or alternate spellings, depending upon your reference) is better known by another name. What is it?
3. What was the previous name of the country now called Zimbabwe?
4. What is a *roman à clef*?
5. Give an example of a *roman à clef*.
6. What is a patten?
   a) a shoe   b) a song   c) a book   d) a typeface
7. "Pale hands I loved, beside the Shalimar" is a line from an old song. What is the Shalimar?
8. For whom was January named?

2

A

aperient

VOCABULARY

Warm-Ups

The warm-up words are taken from the works of Isaac Asimov, Mensa's favorite international vice-president and one of its favorite writers.

1. Medley
   a) interfering or nosy
   b) a type of fruit
   c) motley, varied; music consisting of parts of other pieces
   d) uninteresting, flat, dull
   e) pertaining to the Medes (of the Medes and Persians)
2. Injudicious
   a) against the law
   b) subject to legal penalties
   c) within the walls
   d) lacking sound judgment or discretion
   e) willfully injurious in deed or language

29

3. Deputation
   a) argumentativeness
   b) forcible removal to another place as punishment
   c) showing signs of insanity
   d) dispossession or loss
   e) a group or individual appointed to go on a mission on behalf of another or others

4. Antidote
   a) a brief tale relating something of momentary interest
   b) running in opposite directions around an axis
   c) the curved elevation within the outer rim of the ear
   d) a remedy against poison or an attack of disease
   e) the opposite side of the world

5. Volition
   a) the act of flying
   b) electricity as measured by volts
   c) to volunteer as a worker in an organization
   d) the act, power, or faculty of willing or resolving
   e) speaking with extreme fluency

6. Hazardous
   a) a wordy or roundabout way of speaking
   b) full of risk, dangerous
   c) a line that forms a boundary, or, loosely, any surrounding area
   d) a form of Greek column
   e) a hermit

7. Rigmarole
   a) a form of horse harness
   b) difficult, impossible; a severely tedious task
   c) a succession of incoherent statements, a rambling discourse
   d) a fancy suit of clothing, a costume
   e) a small channel, gutter, or groove
8. Circumscribe
   a) to mark out the limits of, to restrain, to abridge, to draw a line around
   b) carefully, with due regard for common sense
   c) an accent mark over a letter (of Greek origin)
   d) the adjuncts of an action or fact
   e) to write in a circular form, using flowery language
9. Empirical
   a) relating to an empire
   b) based chiefly on experience
   c) pertaining to an adventurous or chivalrous undertaking
   d) related to the study of science
   e) related to New York State
10. Affront
    a) the decorated facade of a building
    b) the false front used sometimes for theatrical performances
    c) an architectural style
    d) an insult, a word or act of intentional disrespect
    e) to disturb, startle, or frighten

11. Petulant
    a) displaying peevish impatience, or (rare) pert, insolent, or rude
    b) full of petals, flowery (obsolete)
    c) pertaining to the keeping of household pets
    d) a species of the genus *Petunia*
    e) an oily hydrocarbon obtained from turpentine
12. Assiduous
    a) a contract of convention between a ruler and his people
    b) a sitting of a legislative body
    c) constant in application, persevering, unremitting
    d) related to the Weights and Measures Bureau
    e) a slender spear or sword (obsolete)
13. Sardonic
    a) bitter, scornful, mocking (usually, of a smile or laugh)
    b) pertaining to a native of Sardinia
    c) pertaining to sardines
    d) pertaining to a sarcophagus
    e) a type of Greek actor in Attica
14. Incongruous
    a) badly fitting, as clothing
    b) disagreeing in type or character, inconsistent
    c) unpleasant, disagreeable, argumentative
    d) unable to be consoled, extremely sad
    e) without money, bankrupt
15. Chronal
    a) on a record or register
    b) of or relating to time (rare)

c) pertaining to many colors
d) staining by immersion in coloring matter
e) pertaining to the larvae of insects

16. Invocation
    a) a dedication to a career
    b) a test designed to measure aptitudes
    c) a calling-upon (usually a deity) for help, a prayer
    d) a liniment
    e) loud shouting

17. Iridescent
    a) recurring at frequent intervals
    b) displaying colors like those of the rainbow
    c) not able to be fixed or repaired
    d) nonnegotiable, as a check that is unsigned
    e) shining blue-white

"Prolix Prose" consists of well-known proverbs or sentences cast in extremely verbose language. Give the original of each sentence disguised in this form:

18. It is considered to be extremely injudicious to investigate with extreme care the oral cavity of an *Equus* which has been provided to one gratis.
19. It is possible to avoid thrice three in repairs if the initial step of repairing a rent or damage is initially made with a filament attached to a sharp, pointed piece of steel with a hole in one end.
20. A superfluity of persons able to prepare edible comestibles sometimes can damage the preparation of a liquid intended for consumption.

A Way with Words

One of the more common complaints that arise in any discussion about intelligence, particularly among critics of IQ tests, is that intelligence is difficult to define. Somehow, the compilers of the *Oxford American Dictionary* (Avon Books, 1980) had no such difficulty:

> **intelligence** *n.* 1. mental ability, the power of learning and understanding . . .

What could be plainer?

Yet, somehow, that definition is perhaps not as simple as it appears. It raises some additional questions: How does one define *power, learning,* and *understanding*? Suppose, for example, Joyce and Harry are given the same set of facts and arrive at the same conclusion. It could be argued that they are of equal intelligence. But what if Joyce arrives at her conclusion in thirty seconds and Harry takes forty-five seconds; does that mean Joyce is a little more intelligent than Harry? (Probably; that's why most IQ tests are timed.)

And how much "learning" are we talking about? Who is the more intelligent, the person who loves mathematics and delves into its seemingly infinite elegances, or the one who "understands" the need for speed and accuracy and therefore learns to operate an electronic calculator with skill and efficiency?

Furthermore, how much of the "power" to learn and understand is dependent upon what one is given—taught, if you will? Several years ago, a television program about IQ tests attempted to demonstrate the effects of environment on IQ scores by describing a test that was administered to a group of inner-city black children. The vocabulary section included the word *lark*. Almost every kid failed to answer because the answers from which they had to choose did not include *car* or *cigarette*, two products with the brand name "Lark" that were being heavily advertised at the time. It hardly seems fair to denigrate the kids' intelligence because circumstances limited their knowledge of birds to pigeons and sparrows.

For decades, IQ test designers have been struggling with the problem of a "culture-free" or "culture-fair" test, one that does not depend on education—a package of learned facts—for successful completion. (They seem to be succeeding; Mensa now has such a test available, and present indications are that it really works, although some psychologists still do not agree that a culture-fair test really exists.) Such tests are really intended as a kind of convenience; you can't take an English or Italian test and translate it into Swahili or Urdu and expect it to be effective; adjustments have to be made. Rather than attempt to fine-tune every test for the linguistic and cultural differences each language requires, it would be better—and easier—to have a culture-free test, a goal that has not yet been completely met.

None of this negates, however, the importance of language in both the determination and the use of intelligence. The "power to learn and understand" is really the ability to communicate, which is the ability to send and receive information and ideas.

Clearly, then, the ability to learn and understand rests solidly on a foundation of words. As the writer and language critic Donna Woolfolk Cross wrote, "In order to understand a thing, you must first be able to describe it." If words are essential to learning and understanding, and if the ability to learn and understand is a mark of intelligence, then it follows that any attempt to measure intelligence must include a measure of one's ability with language, which is why vocabulary and reading comprehension are such important elements of IQ tests.

People who do well on IQ tests are likely to have an affinity for words. That probably explains why Mensans enjoy word games and puzzles so much. Many Mensans are fond of inventing their own word games. Puns, anagrams, double-entendres, spoonerisms, limericks, and other forms of word play abound in the organization's publications and at get-togethers. While some of us rail vigorously against abuses of the

English language, we have no qualms about inventing, as the occasion dictates, our own words. (My own latest creation, arising from the need to describe that which a friend of mine does best, is *hypochondriate.* It's a verb, and whenever I use it, the listener always understands, instantly, what I mean. Also, I will seize this opportunity to record, for posterity, that I was the first person to create a new word of departure, one that will eventually replace *good-bye* or *so long.* In 1980, at the Mensa Annual Gathering in Louisville, Kentucky, I predicted that our progeny will take leave of each other by waving and calling out: "HAND!"—an acronym for the execrable "Have a nice day!" In the South, of course, the word will be HANDY—"Have a nice day, y'all.")

Perhaps it's a reflection of the importance of words that while Mensans have a marked propensity for using them, they don't like to waste them. Most intelligent people have little patience with jargon, corporate or professional gobbledygook, or unnecessary verbiage. Case in point: Hans Frommer, the clever but lazy student with the trick notebook whom we met in the Trivia chapter. Apparently, Hans managed to graduate despite his scholastic lassitude and eventually landed a job with Caterpillar Tractor in Milwaukee. "I was making a monthly report at work," he recalls, an inventory of damaged and broken tools. "It took five hours every month, and I had no idea what use was being made of it. The report was Greek to me, [and] I resented the time it took.

"So, one day," he continues, "I told the boss that I

not only hated to do it, but that it was worthless and he should quit making me do it."

"How do you know it's worthless?" the boss asked somewhat patronizingly.

"Because," Hans replied, "I am six months behind, and nobody has asked for it." The reports were discontinued shortly thereafter.

Understandably, Frommer has become something of an expert in cutting corners to save time. Unable to resist the lure of words, he has compiled an eminently useful pamphlet called *Time Savers* that covers the home, office, kitchen, tax time, and travel. You can have a copy by sending $2 to Lakeshore Publishing, 12360 N. Lakeshore Drive, Mequon, WI 53092.

Rod Vickers, writer, artist, and recently retired entrepreneur of Shawnee, Kansas, used to be regional vice-chairman of American Mensa. Each of Mensa's nine national regions has a number; they also have names. Vickers's domain, encompassing a somewhat amorphous area in the middle of the country, was saddled with the Central Northwest Region, a name that somehow displeased its official representative. Consequently, Vickers was appointed, by the chairman, as a committee of one to come up with a suitable designation for the area. He would regale the American Mensa Committee with his newest offer at each meeting, supporting the proposed name with outrageous, and outrageously verbose, "reports." He finally stopped when one of his names missed becoming official by a single vote, but it is a lasting tribute to his linguistic skills that to this day, Region 7, while still officially known as the Central Northwest Region, is informally and affectionately referred to as "The Great Buffalo Chip Region."

Rod's skill with words is best manifested by his ability to give something resembling credence to what turns out to be, on reflection, exquisitely idiotic. When asked to contribute anecdotes on quick-thinking problem solving for this book, he included the following in a group of several offerings:

> My daddy sent me out when I was just a little tyke and gave me just two bullets, because we needed two squirrels for stew. I wasted one of the bullets shooting a tin can, and knew I was in trouble. My daddy was a believer in corporal punishment, and he acted often on his belief. So I waited until I saw two squirrels sitting side by side, and shot and swung my gun at the same time. Got them both.

He almost got me with that one.

Like Frommer, Vickers enjoys playing and working with words, but has never much liked wasting them. As a child, he quickly hit upon a method for dealing with what was once a common version of that problem, and it is passed along here in the event that there are still teachers who insist that obstreperous students be made penitent the old-fashioned way: "I learned to get through five hundred repetitions of a phrase very quickly," he claims, "if it would fit on one line. By grasping in a line four or five pencils held together by rubber bands, it is possible to write four or five lines of material at a single pass." At this point, it's a little difficult to trust Rod Vickers's credibility, but at least this sounds more plausible than killing two squirrels with a single bullet.

Sometimes words fail, even for those who use them professionally. H. R. "Dick" Horning is now public relations director for the Miami Dolphins, but in 1949 he was fresh out of college and had an urge to see the world. He wound up in Australia, where he landed a job as a reporter. The following year, Dick decided to move to Hong Kong in search of further adventures. Among the passengers on the ship was one Tony Ang, a Chinese seaman who had jumped ship six years earlier, had married an Australian woman, begun raising a family, and become a solid, hard-working citizen. Unfortunately, Tony was on Dick's ship because he was being deported as an illegal—and "undesirable"—alien, in keeping with what was then Australia's "white only" immigration policy. Dick had met Tony earlier, but it was aboard the ship to Hong Kong that he got to know and befriend the luckless ex-sailor, his pregnant wife (who had insisted on accompanying him), and their three sons.

In Hong Kong, Dick joined the *China Mail,* one of the major English-language papers in the Far East. "I did not think of Tony and his family for several months," Dick wrote. Then, almost on impulse, he decided to look up his shipboard companion.

Dick found the Angs in Kowloon, the "mainland" half of Hong Kong, living in a slum and down to the last of their savings. There was no work, and no prospects of any kind for survival in the immediate or long-range future. Tony Ang was seriously and dispassionately considering doing away with himself and his family, rather than have them endure the pain and humiliation of starving to death.

The reporter decided to use his major asset—his words—to help the Angs.

"I . . . worked out an 'open letter' to [the Australian] Immigration Minister . . . from Mrs. Ang," Dick later reported. "She knew little of writing, so I wrote it, freely expressing my outrage." He then got the Hong Kong Bureau Chief of United Press to run the open letter over the international news wires. Australia was in the midst of a heated election campaign, and the local press was eager to publish Mrs. Ang's ghost-written appeal. Within days, Australia had a new government, and the Angs were invited to return.

The family went back home, and Dick Horning lost contact with them. Twenty-five years later, in anticipation of a visit to Australia, he decided to attempt to find Tony Ang once again. This time, however, his method was brilliantly simple. He wrote to the Australian Embassy in Washington and asked for a listing of all the Angs in the Sydney telephone directory. There were four. He wrote to "Anthony Ang," and eventually received a reply from Tony's son, Dennis, who informed him that although his father had since died, all the other Angs were prospering. Dick's words had helped save a family from virtually certain extinction.

But what about that failure of words? Although Dick Horning had found the Angs in the Kowloon slum, he cannot, even now, explain how he did it. "There were no directories of any kind available," he reported, and he knew that the local immigration authorities would have nothing more than a record of the Angs' entry. "Yet," he recalled, "I remember taking a Walla Walla boat to . . . Kowloon, getting on a . . . bus . . . and riding for some three miles and impulsively getting off." He asked a Chinese clerk in a corner tobacconist's shop if he knew where the Chinese man with the Caucasian wife lived. "Oh, yes. I will show you," came the reply, and the clerk walked a block with Dick and pointed out the Angs' shack. "I still do not have the faintest idea how I 'knew' where to find the family in the teeming squatters' huts on the Kowloon side," he claims. "I have resolved the mystery as either a memory failure on my part in recalling how I knew, or by accepting the idea that I really wasn't the one who searched them out, after all."

There are, he suggests, powers beyond those of learning and understanding.

Now try your own power of learning and understanding with the vocabulary quiz, and match wits with Mensa.

Match Wits
with Mensa

VOCABULARY TEST

Time started _____
Time elapsed _____

1. Cupidity
   a) relating to the Roman god of love
   b) adorableness, state of being lovable
   c) inordinate longing or desire to attain possessions
   d) containing traces of copper
   e) sailing in a direction contrary to the wind
2. Enormity
   a) extreme largeness incapable of being measured
   b) monstrous wickedness
   c) extreme volume too loud to be endured
   d) exaggerated rhetoric
   e) state of extreme anger

3. Progenitor
   a) a descendant in the collateral line
   b) someone who is in favor of a particular law
   c) an ancestor or originator
   d) a biological term relating to descent in the male line
   e) a speaker on behalf of a candidate
4. Careen
   a) to run at full speed
   b) a type of tropical fruit
   c) to turn a ship over on one side for cleaning, etc.
   d) an Arabic name
   e) a peal of bells in a tower
5. Nefarious
   a) related to a queen of an Egyptian dynasty
   b) relating to a new form of travel, undertaken for the first time
   c) wicked or villainous
   d) neglectful of assigned duties
   e) a stall holder at a street fair
6. Risible
   a) capable of rising (said of bread dough or cake dough)
   b) motive power for a balloon
   c) having grown to full height, as a plant
   d) capable of producing laughter, laughable
   e) a variety of rice cultivation practiced without water fields
7. Aperient
   a) a type of before-dinner drink
   b) something perceived

    c) an opening in a wall for looking through

    d) laxative

    e) something that appears to be other than it actually is

8. Diaconate

    a) a poison related to belladonna

    b) a learner

    c) a large crown or wreath of leaves

    d) pertaining to the office or rank of deacon, or time when one is a deacon

    e) a type of earth that readily absorbs liquid

9. Obiter

    a) a laudatory description of someone, written after his or her death

    b) an overbite that must be corrected by an orthodontist

    c) by the way, in passing, incidentally

    d) a legal decision against a plaintiff

    e) the file in a newspaper office where obituaries are kept

10. Eponym

    a) one who gives, or is said to give, his or her name to a people, place, or thing

    b) a synonym

    c) the chief officer of an antique Greek state

    d) a variety of restaurant specializing in horsemeat

    e) a mathematical formula for determining liquidity

11. Gibbet

    a) a form of speech that is almost unintelligible

    b) any long-armed ape of the genus *Hylobate*

    c) convex, rounded, protuberant
    d) a gallows
    e) a gibe or taunt

12. Equinoctial
    a) pertaining to horses
    b) pertaining to a state of equal day and night
    c) pertaining to heavy rains
    d) equitable, fair, just
    e) meaning one thing and expressing another

13. Autonomic
    a) a self-propelled automobile
    b) pertaining to a robot
    c) pertaining to a self-regulating calculator
    d) named after oneself
    e) self-governing

14. Erudite
    a) belching
    b) learned, full of knowledge
    c) a form of ornamental stitching
    d) a winding path
    e) of written, as opposed to oral, learning

15. Dour
    a) a variety of tropical fruit
    b) hard or stern, sometimes obstinate
    c) a variety of rock formed under pressure
    d) a Spanish coin of the later Middle Ages, worth
       about 25¢
    e) virtuous, valiant

16. Mendacious
    a) begging
    b) lying, untruthful, false

c) capable of being mended or repaired
d) of high quality, superior in choice
e) generous, giving freely to beggars

"Prolix Prose": These familiar sayings and proverbs have been recast in extremely verbose form. Give the original for each.

17. Individuals who habitually or commonly domicile themselves in habitats containing side portions or roof coverings made of a silicaceous material would be well advised to refrain from casting any size of hard mineral pellets (other than metal).

18. One who habitually seeks repose in the arms of Morpheus at an hour somewhat before that usually considered, and who equally bids farewell to said repose in an equally prompt manner, is reputed to acquire an excellent bodily condition, not to mention an accumulation of pelf and a substantial store of sapience.

19. An individual who possesses neither sapience nor the knowledge to be aware of this lack will rapidly find that his stock of worldly goods, chiefly in the form of coin of the realm, has vanished.

20. It is considered to be extremely injudicious to attempt the feat of transferring from one member of the genus *Equus* to another member of the same genus while engaged in the process of being transported through or over a moving body of H_2O.

3
LOGIC,

J 31, A 31,

S 30, O 31,

N 30, ?

REASONING, &
MATHEMATICS

Warm-Ups

1. If 8-22-5-22-13 equates to seven, and 7-4-12 makes two, how would you write ten?
2. The following word square makes a fractured proverb. If you start at the correct letter and proceed in any direction, one letter at a time, using each letter only once, you will find the fractured proverb. (Hint: Start with the first letter of the alphabet.)

N E L O A
S S T L L
T E G T H
R S I L A
E T T G T

3. Back in the days when there were maidens around who could see unicorns, two of the young ladies passed a field in which some unicorns and some rams were prancing about. One young lady

remarked that she could see a total of forty-four horns. The other young lady remarked that there were twice as many unicorns as rams. How many unicorns and how many rams were there?

4. All gloops have hard skins; all slooms have soft pits; all slurps are juicy. Some gloops are blue; some slooms are red; all slurps are yellow. Some gloops are yellow, and some slooms are yellow. Which of the following do I know about a sloom, based on the above information?
   a) It is red.
   b) It is juicy.
   c) It has a hard skin.
   d) It has a soft pit.
   e) none of the above

5. Which of the lettered objects comes next in the following series?

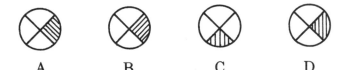

A   B   C   D

6. Mary won't eat fish or spinach; Sally won't eat fish or green beans; Steve won't eat shrimp or potatoes; Alice won't eat beef or tomatoes; Jim won't eat fish or tomatoes. If you are willing to give such a bunch of fussy eaters a dinner party,

which items from the following list can you serve?

green beans   creamed codfish   roast beef
roast chicken   celery   lettuce

7. A girl decides to take a long walk in the country and visit a friend on the way. She walks at a steady pace of 2½ miles per hour. She spends 4 hours walking over to her friend's house; she has a cup of coffee and a sandwich and talks to her friend, all of which occupies an hour, and then her friend runs her home in the car, over some rough road, at 20 miles per hour. She gets home at 2:30 in the afternoon. When did she leave her house in the morning?

8. The same five letters, if rearranged, can be used to fill in both sets of blanks in the following sentence, to make a sensible sentence:
Each bank - - - - - that it has the very best plan for each - - - - -.

9. What is the opposite of the following scrambled word?
RUECTLY

10. Jake's jalopy uses 10 gallons of gas for a trip of 150 miles. Sam's speedster gets half the mileage that Jake's does. Hal's heap uses 10 gallons for two-thirds the distance that Jake can go with his jalopy. How many gallons does Hal need to go 250 miles?

11. Under certain special circumstances, a peach

costs 20¢, a banana costs 30¢, and a grapefruit costs 40¢. How much will a pear cost under the same circumstances?

12. "A bird in the hand is worth two in the bush" means approximately the same as:
    a) If you go hunting and shoot nothing, you have wasted your day.
    b) Better a small gain you are certain of than a larger gain that you may not be able to obtain.
    c) If you raise chickens, you will always have one to cook for dinner.
    d) It is extremely difficult to catch a bird by hand, so you should be very careful with one when you do catch it.
    e) Be careful what you do with birds; they can be dangerous.

13. If *through* and *threw* are pronounced the same, cross out all of the even numbers in the line below. If they are not, cross out all of the odd numbers. If Albany is the capital of New York State, add all of the numbers that are left. If it is not, multiply all of the numbers left. Your answer?

    1 2 3 1 2 3 1 2 3 1 2 3

14. What comes next in the following series?
    J 31   A 31   S 30   O 31   N 30   _____

15. Which of the numbered figures best completes the following series?

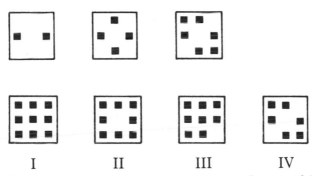

I   II   III   IV

16. You are an employment manager in that mythical land where there are only liars and truthtellers. An applicant comes in to see you and appears to be sincere. He tells you that the next interviewee has admitted to him that she is a liar. Is he lying or telling the truth?

17. If forty pizza bakers can bake twenty pizzas in two hours, how many hours will it take two pizza bakers to make ten pizzas?

18. What is the next letter in the following sequence?
    A D G K O T ____

Rebuses: Mensa members have recently taken up the type of puzzle known as a rebus. These are pictures that suggest a familiar word or phrase. They are not always what they seem, sometimes contain puns, and are usually best solved by wiping your mind clear and thinking of the exact description of the drawing. Try these for practice:

Practice rebus:

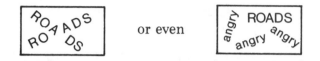 or even

The answer, of course, is crossroads. Now try the next two.

19.

20.

It's Only Logical

Many people confuse education and intelligence. "He must be pretty smart; look at all the degrees he has!" In fact, the multidegreed "genius" may be clever only at avoiding honest work by extending his education.

An education is nothing more than a set of tools, equipment with which to accomplish whatever one needs or wants to do. Intelligence, on the other hand, is the ability to select from those tools the ones that can help perform a task with the greatest efficacy. When it comes to practical problem solving, education isn't much good without intelligence, and vice versa.

In a sense, logical thinking can be defined as taking a set of known facts, adding some other known facts, and arraying them in such a way as to come to some inevitable conclusion. It's what computer and management types refer to as the "if-then" technique. Mathematics represents the ultimate in logical thinking because, except in its most obscure and abstract forms, mathematics is always the inevitable result of a logical sequence of events.

Admittedly, virtually all of the stories in this book that show how Mensans use their intelligence in real problem-solving situations can be classified as logical thinking (except for those in the epilogue, which are obviously illogical thinking). The ones in this section, however, seemed to be particularly fine examples of how relatively simple logic—the combining of facts from existing circumstances and from "stored" information—can find solutions to problems. "Reasoning" is really a logic technique. It involves laying out the available information, mentally or on paper, and arranging it and rearranging it until a workable or "logical" sequence falls into place. Depending upon the individual and the problem, that can sometimes be a matter of seconds—or days.

For example, what do you do when you have to move a giant tree? If you're Phyllis J. Dutrow, of St. Petersburg, Florida, you ponder the situation for a while and determine the facts, and a reasonable solution soon presents itself.

For reasons she does not explain, Phyllis had the need to extricate a cabbage palm from its moorings. The tree in question was, she said, "about as big around as a fat man's thigh and . . . in a hole more than knee-deep." (Those measurements are perhaps a trifle imprecise, but you get the idea.)

"These things," she continued, "have bulbous ends with a multitude of roots, each of which is roughly the diameter of a lead pencil, that extend several feet.

After turning the tree in the hole and reaching under to sever [the roots] with a left-handed tomato knife . . . I determined that the palm no longer was glued, screwed, and through-bolted to the far side of the earth. Problem: how to get it out of the hole, which was alongside the sidewalk." (Somewhere in that paragraph, Phyllis mentioned that the left-handed tomato knife had cut far more palm and palmetto roots than it had left-handed tomatoes, but I decided not to include that information because it's irrelevant.)

First she shoved the tree to one side. Then she shoveled a few inches of sandy soil into the gap on the other side. Next she shoved the tree over to the side with the soil, which was now four inches higher, leaving a gap on the first side, which she promptly filled with more soil, and then pushed the tree onto *it*. "Alternately fighting the palm from side to side as I refilled the hole a few inches at a time, I got the monstrosity to sidewalk level. I tipped it flat on the walk and rolled it more than a hundred feet to the dumpster. I didn't want to tarry among the fire ants to figure how to get it in there," she said, so once again she resorted to quick thinking and logic: "I had a couple of neighbors lift it in. . . . "

(Phyllis's problem—and her clever solution—bears an uncanny resemblance to the dilemma faced by the good people of Pisa, whose famous tower, they believed, was beginning to lean a little more than somewhat and was in danger of eventually rendering itself horizontal. A Mensa member not only solved the problem but was able to reassure the Pisans that there really was no problem. I've decided that a story about the Leaning Tower of Pisa belongs in the Culture section, where you'll find it.)

The late Jimmy Durante could always count on a laugh with his famous line "Don't raise the bridge, men; lower the water." Perry Oliver, of St. Joseph, Missouri, used simple logic to apply that basic principle to an annoying, and seemingly insoluble, difficulty.

"When Wavalea [Mrs. Oliver] and I first moved here, I had a problem with her about a clothesline," Perry wrote. "She kept complaining about it sagging. I kept tightening it, and it kept sagging. She wouldn't believe that a clothesline couldn't be straight." The fact is, however, that it couldn't because the weight of the wet clothes would invariably stretch the line and cause it to sag, despite the fact that it was, according to Perry, "as high on the poles as it would go."

"She wanted taller poles, which I didn't have," he continued. Perry's personality and life-style do not lend themselves to acquiring new things when the old things are still serviceable, so new, longer poles were never a serious consideration. "My tractor and grader blade were handy," he said with his usual practical succinctness, "so I lowered the ground between the poles."

Perhaps nowhere does the combination of quick thinking and a little knowledge of basic science prove more effective than in combating the elements. The tale is reported by Jerry Salny:

"When I attended the Massachusetts Institute of Technology," Jerry writes, "I lived in a fraternity house. . . . Six of us, on double-decker beds, slept in each room. The rooms were kept dark and quiet, and with the windows open." Unfortunately, the climate of Boston is given to sudden changes, with what Jerry describes as "horizontal rain, snow, and sleet." Also unfortunately, such changes tend to occur in the middle of the night, wreaking havoc on a frat dormitory with an open window and a half-dozen students, each feigning sleep in the hope that someone else would get up and shut the window. Soon, however, one of the occupants hit upon a solution.

"First," relates Jerry, "he cut the sash cords so the window would fall by itself. Then he propped open the window with a stick. Next he put a sugar cube under the bottom end of the stick. When the rains came, the sugar cube melted away, the stick fell out, and down came the window." Of course, the stick was tied to a string to prevent its plummeting four stories to the ground.

"Now," concludes Jerry, "in my opinion, the fact that Dr. Richard P. Feynman [the perpetrator of the just-described ingenuity], Tolman Professor of theoretical physics at Caltech, later won the Nobel Prize in physics, simply pales into insignificance compared to his having figured out how to close a window on a cold Boston night without getting out of a warm bed."

If that is so, then there is no telling to what heights Morris Berwick, of New Orleans, may yet ascend. His tale of logical thinking is my personal favorite among all the stories sent in by Mensans. Let him tell it in his own words:

"I spent a year in Wichita, Kansas, in a mobile

home. During my tenure, I got to see both trees. I also got to see nineteen degrees below zero, with thirty-mile-an-hour winds.

"The aluminum awning attached to the end of the trailer had tried valiantly to resist the wind, but to no avail. Finally, it separated, all the way along one edge, and became a battering ram that tried to destroy its own now-useless mounting.

"This happened in late evening. Since it was adjacent to our bedroom window, sleeping was fitful, at best. I had considered going outside to pop-rivet it together, but just opening the door was enough to convince me that my hands wouldn't last long enough.

"After a very light and troubled sleep, I awoke at 1 A.M. with the solution. I took a bath towel and soaked it with water. I then ran outside and threw the towel across the break. The towel froze solid. Fiberglass and resin could not have made a better patch.

"Two quiet weeks later, when the temperature rose above freezing, the patch fell off. I then made a permanent repair in relative comfort."

And eventually got the hell out of Kansas and moved to New Orleans.

In each of these instances, the problem solvers examined the circumstances, matched them up with appropriate bits of information, and logically arrived at a conclusion. Now it's your turn to do the same. On to the next quiz.

Match Wits with Mensa

LOGIC,
REASONING, AND
MATHEMATICS TEST

Time started _____

Time elapsed _____

1. All of my grandchildren are under eighteen. All of my granddaughters are very beautiful. All of my grandchildren have red hair and blue eyes. My oldest grandchild has very long red hair. The legal voting age is twenty-one. Which of the following statement(s) can be proved by the information given?
   a) My oldest grandchild may not yet vote.
   b) My oldest grandchild is a beautiful girl.
   c) My youngest grandchild may not yet drive legally.
   d) My youngest grandchild has short red hair.

2. The following word square is a takeoff on a well-known proverb. If you start at the correct letter and move in any direction, one letter at a time, you will find the slightly fractured proverb. Each letter may be used only once. There is a dash in the sentence, which is included also. (Hint: The sentence starts with THE.)

N	R	A	E	E	T
E	L	T	E	G	S
T	Y	H	R	M	W
A	E	W	O	A	H
S	—	E	S	T	I
E	V	R	E	D	T

3. You have twenty-four socks in a drawer, six each of brown, black, white, and red. How many socks must you take out of the drawer, without looking, to be sure of having a matched pair (of any color)?

4. A man walks to his friend's house at 2 miles per hour. He spends an hour eating lunch, and then rides home on his friend's bicycle, five times faster than he walked. The distance to his friend's house is 10 miles. At what time must he leave home in order to complete the round trip and return home by 4 P.M.?

5. Which of the lettered words could logically come next in the following sequence?

APE BIRD CAN DIG EAT
a) MAN b) HAT c) CAR d) SEA
e) FIG

6. Sally goes out on a shopping expedition for some new clothes. She buys a coat, a skirt, and a scarf. The coat costs twice as much as the skirt, and ten times as much as the scarf. She starts out with $180 and comes back with $20. What was the cost of each item of clothing?

7. The same five letters, if rearranged, will make two different words to fill in the blanks in the following sentence:

Those who wish to be good - - - - -, it is said, must have excellent memories, otherwise their stories run off the - - - - -.

8. "Birds of a feather flock together" means approximately the same as:
   a) All songbirds stick together.
   b) Feathered birds get along well.
   c) People tend to congregate with others like themselves.
   d) If you see a lot of birds together, they will probably be the same color.
   e) Birds without feathers are not accepted by birds with feathers.

9. In the square at the left below, a rule of arithmetic has been followed which applies both across and down. Find the rule and supply the missing number to the square at the right.

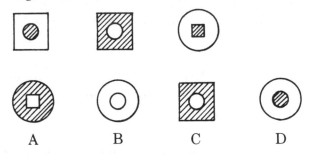

15	3	5		24	4	6
5	1	5		6	1	—
3	3	1		4	4	1

10. Which lettered figure best completes the following series?

A   B   C   D

11. So you're good at Roman numerals. Here's a different use for them:
CXV = SIX; CMLMI = SEVEN.
What does LMXI equal?

12. Each of the letters in the puzzle below represents a number. The answer will be the same numbers, reversed, as shown. Fill in the numbers.

$$\begin{array}{r} ABCD \\ \times 9 \\ \hline DCBA \end{array}$$

13. What word is the opposite of the following scrambled word?
CWCADORIE

14. Which number in the following series of numbers is least like the others?
1 3 5 7 11 13 15

15. Which of the following scrambled words is the

"odd man out" when the words are unscrambled?
CGHICOA   TTOOORN   IMMIA
CPOEHNANGE

16. Four couples are going to the movies. Each row holds eight seats. Betty and Jim don't want to sit next to Alice and Tom, and Alice and Tom don't want to sit next to Gertrude and Bill. On the other hand, Sally and Bob don't want to sit next to Betty and Jim. How can the couples arrange themselves so that they all sit where they would like?

17. Which of the lettered designs best completes the sequence in the first row?

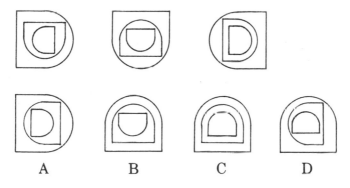

A   B   C   D

18. If *diaphanous* and *sheer* do not have the same meaning, cross out all the 9's in the line below. If they do, cross out all the 6's. If *slough* and *cough* are pronounced the same, multiply the number of 4's by 6. If not, add up all of the non-crossed-out numbers and multiply by 4.
9 4 6 4 9 4 6 9 4 6 9

19. The price of an article is cut 40% for a sale. When

the sale is over, the store owner wants to bring the price back up to the original selling price. What percentage of the sale price must be added to that sale price to bring the price back up to the original selling price?

20. All readers of this book are clever. Some readers of this book are Mensa members or will join Mensa. Some readers of this book like to solve puzzles. Therefore, which of the following statements can be justified by the information just given?

a) All readers of this book like to solve puzzles; otherwise they wouldn't be reading this puzzle.

b) All clever readers will join Mensa.

c) Some Mensa members like to solve puzzles.

d) none of these from the information given

4

ANALOGIES

Warm-Ups

1. The Panama Canal is to the Atlantic and Pacific as the Suez Canal is to _____ and _____ .
2. *p* and *q* are to *g* as *b* and *d* are to:
   a) *y*   b) *g*   c) *j*   d) *a*
3. *Bankrupt* is to *solvent* as *grouped* is to:
   a) discrete   b) assembled   c) massed
   d) schooled
4. D. W. Griffith is to the motion picture as Henry Fielding is to the:
   a) opera   b) story novel   c) stage play
   d) epic poetry
5. The Empire State Building is to New York City as the Sears Tower is to _____ .
6. Lockheed is to airplanes as Stutz is to _____ .
7. Paul Bunyan is to Babe as Don Quixote is to _____ .
8.

is to   as   is to:

A   B   C   D

9. President William Howard Taft is to Senator Robert A. Taft as President John Adams is to _____ .
10. Piccard is to ballooning as Cousteau is to _____ .
11. *Humoresque* is to Dvořàk as *Minute Waltz* is to _____ .
12. William H. Porter is to O. Henry as Eric Blair is to _____ .
13. The Underground is to London as the Metropolitain is to _____ .
14. Trenton is to New Jersey as Pierre is to _____ .
15. Leo Durocher is to baseball as Knute Rockne is to _____ .
16.

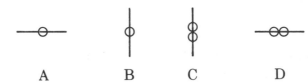

A   B   C   D

17. A scepter is to a king as a mitre is to a _____ .
18. The Packers are to Green Bay as the Chiefs are to _____ .
19. The Ilyushin is to Russia as the Hawker-Siddeley 748 is to _____ .

20. The telephone is to Alexander Graham Bell as the mercury thermometer is to _____ .
21. The Japanese language is to Japan as Tagalog is to _____ .
22. Chosen is to Korea as the Gold Coast is to _____ .
23. The Nile is to Egypt as the Volga is to _____ .
24. The New York Cosmos are to soccer as the Toronto Maple Leafs are to _____ .
25. Henry Fonda is to *On Golden Pond* as Rex Harrison is to _____ .

Look at It This Way...

Solving analogy puzzles involves the ability to see relationships between sets of situations that may not be immediately obvious. To be sure, that often requires some special knowledge, such as a nodding acquaintance with, say, history or botany. (Nothing heavy; any good shopper probably knows enough about botany—which is to say, fruits and vegetables—to handle such questions on a typical analogy quiz.)

But apart from the usual and acknowledged fun and diversion in working such questions, is there any real value to them? Would I be asking that question if the answer were no?

The ability to understand that a given situation is analogous to some other situation, and, in so doing, to solve a problem, is something of a talent, but it is also a skill that can be developed with practice. Rather than lecture you on how using analogies can enhance your understanding as it frees your thinking, let's attempt to prove it by example.

This is a true story, but I won't use real names, to avoid the risk of offending someone—particularly someone who authorizes the issuance of a check to me on a regular basis. I write a column, for a fortnightly international business publication, on computers and telecommunications. For the most part, I'm expected to generate my own stories; technology being what it is these days, there's hardly a scarcity of material. The catch, however, is that in every case, there has to be a direct tie-in with international trade and/or transportation.

One day, I received a press release in the mail, sent to me not by the public relations firm promoting the company, but by the associate editor with whom I've been dealing these many months. It described a computer service that made it possible to settle automobile collision claims more promptly and fairly through computerized analyses of the damages as recorded by inspectors. Normally, it might merit half a column in some other context; to me, it was worthless. Nevertheless, attached to the press release was a note from the associate editor, whom I'll call Denise: "Call me when you get this."

I dutifully followed instructions. "A so-so story," I said, "but not for me."

"Harold wants you to write it up for your column," Denise said. Harold is the executive editor. My "column" runs a minimum of 1,500 words. "That's idiotic," I replied. "What does this have to do with international trade and transportation?"

"I asked Harold that same question," she said.

"What did he say?"

"He said they have automobile collisions in Europe, too."

"That's ridiculous," I replied, trying desperately— and not succeeding—to maintain a modicum of composure. "We're talking export-import here; we're talking international banking here; we're talking cargo ships and planes and ports and piers and marine insurance here. We're not talking European automobile collisions!"

"Harold wants the story," she said, through what I was certain were clenched teeth. The conversation was over.

For the next three days, that press release sat on my desk. Approximately once every hour, I picked it up, read it through, and angrily tossed it aside.

Then, literally in the middle of the night, inspiration struck. On the fourth day, I telephoned the public relations man who had issued the release and said, "I have two questions for your client. Their answers will determine whether I do this story. One: If your system works for damaged cars, can it be adapted to work for damaged ships, trucks, and cargo containers? Two: If the answer to the first question is yes, would your client entertain offering to license the system to overseas computer service organizations?" Happily, he came back with the right answers, resulting in a highly successful 2,000-word story that made everybody happy, especially the public relations man and Harold. I think even the readers enjoyed it.

They don't call me the King of Transitions for nothing.

Norris Swan, of Hastings, Nebraska, can also claim

credit for being able to turn over a situation in his mind and look at it from different angles. One evening, he arrived late for a meeting of a church carpet committee to which he belonged. His colleagues had determined that the carpeted center aisle of the church had to be widened. Norris explains:

"The seats consisted of benches, arranged in five sections, with a center aisle, two middle aisles, and two outside aisles. The benches were curved as necessary for the arch. Those in the front were short, and the benches got progressively longer toward the back.

"The committee was in the process of arranging to have the two center sections of benches taken to a millworks to have one-foot sections taken out of each end and then have the end arms reattached. (Two-foot sections could not be taken from one end because that would destroy the curvature of the benches.)

" . . . I asked why we didn't simply take out the back row of benches and move all the other benches back one row. If we needed the additional seating of the lost row in front, we could make a small section for these front seats from one or two of the back benches. It was decided we could use the space up front for a better purpose than seating, so that is what we did." To the never-ending gratitude, probably, of the budget committee.

I have no doubt that somewhere in this world, a group of financially concerned citizens are equally grateful to James A. Harris, of Redondo Beach, California. His story is reproduced here verbatim:

"Working under an oil derrick with a ten-ton rating, I was asked to calculate the safety margin when lifting an eight-ton load. While drawing vector-force diagrams for each cable (like they teach us in school) it became obvious that the strain on the derrick was simply the weight of the load plus the strain on the line that went to the hoisting drum. The hoist was rigged by block-and-tackle to give a mechanical advantage of 4:1, so the derrick load was 8 + (8/4) or no margin at all. We rerigged to give the hoist drum a 6:1 advantage, thereby reducing the strain to 9.33 tons."

I sincerely hope that's all perfectly clear to you, because it makes no sense to me whatsoever. If you have any questions, you can send them to me. I'll forward them to Harris.

When I was a kid, a popular expression was "If the mountain will not come to Mohammed, then Mohammed must go to the mountain." As a schoolteacher, James T. Hart, of Napa, California, applied that simple notion in a masterful manner.

"I was faced with the problem of teaching math to intermediate-grade children who were at varying levels of proficiency in mastering their basic facts as well as the algorithms, fractions, decimals, etcetera," Jim told us. Traditionally, a teacher presents a mathematical concept to the entire class, and all the students work on the accompanying exercises, whether they're capable of doing them or not. That encourages, as some of us know all too well, a sink-or-swim situation, which in turn leads to disruptive behavior and other discipline problems.

"I set out to develop an individualized prescriptive math program that allowed me to chart each student's programs at whatever proficiency level he could achieve. . . . I admittedly spent a lot of hours and used a lot of other people's creations to put my program together. When I was finished, each child had an individual contract assignment spanning a week or so of work, which was self-corrected. Upon completion of each contract, the child would take a self-corrected posttest. If the child passed that test with 80 percent or higher, I would administer the final posttest that I corrected. . . .

"Some of the more motivated students were doing trigonometry by the time they were out of the sixth grade, and almost without exception, my students scored well above state and national averages on the standardized tests the district required all schools to administer."

Unless you've been a math student in California, you may not have benefited from Jim Hart's innovative teaching methods. But if you've ever had a tooth drilled, you certainly have benefited from John F. Johnson's ability to draw an analogy between an automobile factory and the human mouth.

"In 1947," writes Dr. Johnson (you knew he'd turn out to be a dentist, didn't you?), "I worked on drill presses for the Ford Motor Co. and found it was essential to keep the coolant flowing on the drills or they would get hot and be ruined.

"In 1948, as a junior in the School of Dentistry, University of Michigan, I started drilling on the teeth of live people, in the required way, with no coolant. A former dean had convinced the world of dentistry that drilling should be done dry. But here the drill got hot, too, and it hurt," he said, in a masterpiece of understatement.

Shortly after graduation, the Korean "conflict" began, and Dr. Johnson wound up practicing dentistry at Fort Knox, Kentucky. "A friend of mine was in charge of the machine shop," Dr. Johnson recalls, so he had the friend rig a dental drill that would use diamond burrs instead of steel, and at very high speeds—cooled by water. A tank-type vacuum cleaner was adapted to suck up the overflow.

By 1958, a former professor whom Dr. Johnson identifies only as "Dr. Peyton" promoted the teaching of the water-cooled method, and soon, he claims, "high-speed dentistry became the hot issue." (I refuse to believe the pun is unintentional.) "Within about ten more years," he adds, "practically all dentists were using high-speed techniques and high-volume suction, and patients were beginning to realize that dentistry had gotten out of the dark ages. . . ."

I was intrigued by the notion that if the young John Johnson had not worked in an automobile plant, we would still be screaming in terror every time we looked at a dentist, so I asked him, pointblank: "Are you, in fact, the 'inventor' of water-cooled dental drilling?"

His reply: "It is possible somebody would contest the statement that I am the 'inventor' of water-cooled drilling. People still disagree as to who should be named the inventor of the airplane, and evolution, and the telephone. But the chance that anyone will dispute the sequence I gave as to the development of water cooling is slim. . . . I used it exclusively from the day I started practice, at Fort Knox, Kentucky, and about sixty dentists there had it used on them. They were from all over the country, and it was new. I doubt if there will be any real disagreement on the start of the use of the system." That's good enough for me. And for my dentist. And it should be for you, too.

It has long been my contention that being able to see how things are somewhat the same makes it possible to see the same things a little differently. Betty Kubicek, of Milwaukee, helped prove the point. One Sunday morning, her father had undertaken the task of repairing his favorite clock at the kitchen table. "The morning ended," she wrote, "with the clock's successful dismantling, and by late afternoon, he had it all put together again and ticking perfectly." She was, therefore, somewhat surprised, when preparing to set the table for dinner, to find Dad still at it and, as she described it, "the clock being prepped for open-heart surgery once again. 'I thought it was fixed,' I said. 'It is,' he muttered. 'Only thing is, the damned thing only runs upside-down!' "

At this point, Betty did a terrible thing. She picked up the clock and turned the printed-paper clock face a full 180 degrees so that the 6 was where the 12 had been and vice versa.

"I'm not sure it was really so clever of me," she now

says. "The clock worked great, but Dad didn't speak to me all through dinner."

Apparently, the ability to stretch one's mind by making analogies works best when accompanied by a healthy portion of diplomacy.

You won't need any diplomacy for the next quiz. Nobody's looking; go ahead and show yourself how good you are.

Match Wits
with Mensa

ANALOGIES TEST

Time started ⎯⎯⎯⎯⎯⎯
Time elapsed ⎯⎯⎯⎯⎯⎯

Analogies are relationships. Some relationships are obvious and some are more subtle. Each of the following questions gives one set of relationships and asks you to fill in the missing relationship of the second pair.

1. Roquefort is to France as Liederkranz is to ⎯⎯⎯⎯⎯ .
2. Triskaedekophobia is to the fear of the number 13 as photophobia is to ⎯⎯⎯⎯⎯ .
   (Hint: It is *not* the fear of being snapped by a photographer in a less-than-desirable situation.)
3. Florentine is to spinach as Parmentier is to ⎯⎯⎯⎯⎯ .
4. LAX is to Los Angeles, California, as EWR is to ⎯⎯⎯⎯⎯ .

87

5. Drawer is to reward as Anna is to _____ .
6. Wolves are to pack as whales are to _____ .
7. The musical *Oklahoma!* is to *Green Grow the Lilacs* as *My Fair Lady* is to _____ .
8. Mary Ann Evans is to George Eliot as Charles L. Dodgson is to _____ .
9. Cape Horn is to South America as Cape of Good Hope is to _____ .
10. Conforming is to orthodox as nonconforming is to _____ .
11. Enervate is to weaken as energize is to:
    a) strengthen   b) weaken   c) improve
    d) add new batteries
12. Franc is to France as dirham is to _____ .
13.

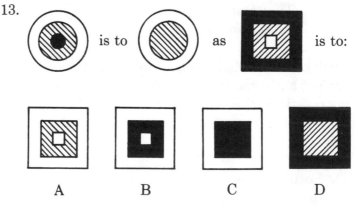

A   B   C   D

14. Profitable is to remunerative as fallacious is to:
    a) fraudulent   b) reward   c) possible
    d) proverbial   e) carnivorous
15. Concealed is to secret as overt is to:
    a) occult   b) science   c) ancient
    d) revealed   e) none of these

16.

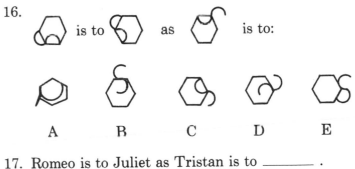

A   B   C   D   E

17. Romeo is to Juliet as Tristan is to _____ .
18. Watergate is to Nixon as Teapot Dome is to:
    a) Harding   b) Wilson   c) Truman
    d) Jackson   e) Pierce
19. Golden Gate Bridge is to San Francisco Bay as
    Bridge of Sighs is to:
    a) Venice   b) Paris   c) London
    d) New York   e) Istanbul
20. Linen is to flax as glass is to:
    a) acrylic   b) polyester   c) sand   d) water
    e) stone
21. McKinley is to Roosevelt as Roosevelt is to:
    a) Carter   b) Taft   c) Truman   d) Garner
    e) Barkley
22. Constantinople is to Istanbul as Ceylon is to:
    a) Zimbabwe   b) Sri Lanka   c) Malagasi
    d) Malawi   e) Papeete
23. Narcissus is to flower as Arethusa is to:
    a) tree   b) rose   c) stone   d) fountain
    e) mountain
24. Gray is to anatomy as Ali-Bab is to:
    a) dressmaking   b) cooking   c) grammar
    d) history   e) newswriting

25. 3 is to 5, 8, 9 as 4 is to:
    a) 1, 11, 14   b) 2, 3, 6   c) 1, 13, 16
    d) 16, 15, 9   e) 1, 3, 6

5

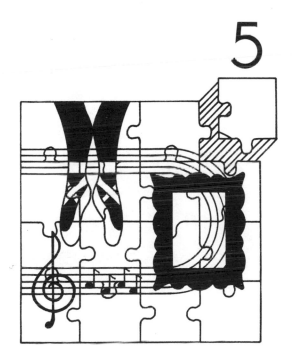

CULTURE

Warm-Ups

ART

1. Many people know that the world-famous paintings of Michelangelo, including *The Last Judgment,* are in the Sistine Chapel at the Vatican. Why is it called the Sistine Chapel?
2. The series of famous Japanese pictures called *The Floating World* is about a specific subject. What aspect of Japanese life is depicted?
3. François Auguste René Rodin was a famous French artist. For what field of art is he best known?
4. Raphael (Raffaello Sanzio) was best known as a painter, but he had another artistic talent. What was it?
5. Anyone who takes a tour of St. Paul's Cathedral in London, or of many English country houses, will hear the name Grinling Gibbons. Who was he?

MUSIC

1. Why were the Gilbert and Sullivan operetta players known as Savoyards?
2. What is the national anthem of France?
3. In what country does the opera *Aida* take place?
4. Have any famous waltzes come down to us from Elizabethan days?
5. In what city in the United States is jazz supposed to have originated?

GREEK AND ROMAN CULTURE

1. They were known as Jupiter and Juno in Rome. What were they called in Greece?
2. Who were the twins that were popularly believed to have founded Rome?
3. What was the famous device used to enter the city of Troy by trickery?
4. Sappho was a famous Greek poet whose name is specifically associated with the island where she lived. What was the island?
5. What was the usual occupation of the Fates?

LITERATURE

1. What was the chief characteristic of female roles on the stage in Shakespeare's time?
2. What are the collected works of the great early Norse poets and writers called?
3. John Milton suffered from a major handicap for a

large portion of his adult life. He refers to it in many of his works. What was it?

4. What was Dante (Durante) Alighieri's greatest work?
5. What is the subject of Tennyson's series of poems, *Idylls of the King*?

ETIQUETTE

1. What shape of glass is preferred by many connoisseurs for champagne?
2. If you are introducing an old man to a young woman, who is introduced to whom?
3. When does a lady take off her gloves to shake hands?
4. What should be printed on a thank-you card for wedding presents?
5. When may you eat lamb chop bones (after the meat is trimmed) with your fingers?

More Than Yogurt

One person's "culture" is easily another's boredom, which makes Culture with a capital *C* nearly as difficult to define as intelligence. Even my favorite dictionary, the *Oxford American,* noted for its terseness, lists five definitions for the word, ending with "a quantity of bacteria grown for study." (Yogurt is a cultured milk product.)

And, as with the definition for intelligence, the first dictionary entry, while seemingly clear and direct, raises several unanswerable questions: "the appreciation and understanding of literature, arts, music, etc." Fine. Now all we have to do is define *appreciation, understanding, literature, arts,* and *music,* and it becomes crystal-clear.

Let's see . . . music. I love Mozart and abhor Wagner. I can't stand any of the composers who follow Tchaikovsky—except for Rachmaninoff. (Within seconds after that last sentence was written, *Rhapsody in*

Blue began issuing forth from the radio. Add Gershwin.) I can listen to Dixieland far into the night (and have), but I can't tolerate acid rock. Disco gives me a headache, but reggae is strangely soothing—almost hypnotic. Indian music hurts my ears; Greek music makes me passionate.

There's more, but by now you're probably bored to stupefaction, so I won't burden you with my views on "arts," which would have to begin with categories—what kind of "arts" do we mean? Performing? "Fine"? Graphic? Conversational? Medical? Why list literature separately? Isn't that one of the "arts"?

As for the "etc." in that definition, it's a prime example of a lexicographical cop-out. But we can't really blame the dictionary writers; they used the "etc." catchall because "culture" is in itself a catchall; in terms of appreciation and understanding of the so-called finer things in life, it can mean virtually anything you want it to mean. (In Mensa, for example, it often means the understanding and appreciation of food and drink.)

The vague definition of culture presented some problems, as you can well imagine, in selecting questions for the quiz that follows and in choosing anecdotes to describe how Mensans apply their intelligence to solving problems within a cultural context. We decided to cop out, too: The questions have to do with the traditional concepts of culture, which may differ a little from yours and mine, but which most people tend to associate with the word. As for the anecdotes, their inclusion in this section is somewhat arbitrary, but so are all the other selections and placements in this book.

One of the best places to observe Mensans' notions of culture is a kind of convention called a Regional Gathering (or, more familiarly, RG, perhaps because that's so easy to pronounce as "orgy"). In 1982 there were over thirty-five RGs, all over the country.

RG workshops and activities can range from lessons in massage to discussions of the criminal justice system, from exhibitions of art by Mensans to regional history, from trivia contests to classical and X-rated films, and on and on. RG hospitality suites are always filled to capacity with people actively engaged in conversations as varied as the participants themselves. Those conversations tend to be liberally lubricated with many gallons of free-flowing domestic wine and beer (yet another tribute to Mensans' eclectic notion of "culture") and snacks. Not everyone, however, partakes of the free refreshments.

Ann-Marie Dobbs, American Mensa's groups officer (whose responsibility is overseeing national special-interest groups), likes a little wine as well as the next RG devotee, but she also likes a change occasionally.

"A friend and I took our own champagne to an RG because we were tired of hospitality-suite Chablis," she writes. "Once we got the wine chilled, we realized we couldn't walk around with glasses of champagne unless we were prepared to share. Since we had brought only a half-dozen bottles, and they had to last an entire weekend, that was out of the question. My friend dug two empty beer cans out of a trash can. We washed them out, filled them with champagne, and aroused no more comment than 'I didn't know you were a beer drinker.' "

For many Mensans, the aforementioned "appreciation and understanding" begins at an early age, so it's not surprising that they encourage a love of "culture" in their children. Linda Kelso, of Portland, Oregon, uses her spare time to edit *Interloc,* a kind of "house organ" for Mensa officers and others involved in the inner workings of the society. Several years ago, she solved a parent-child problem with a little culture.

"My second child was born when the first was just eighteen months old," Linda relates. "What to do with child No. 1 while child No. 2 nursed presented a dilemma because No. 1 already had the how-to-get-attention routine down pretty well.

"No. 1 loved to have books read to him, so nursing time became reading time. No. 1 would sit quietly beside me with no jealousy or interruption."

She added that the older child is now a National Merit Scholarship winner and a Mensan—no small feat for a kid who has to go through life with a name like No. 1.

Clare Neuman, of Allendale, New Jersey, recognizes that among the attributes of a "cultured" person is the ability to be a good host. "People say entertaining is difficult. They're wrong," she insists. "All it takes is some smarts. When people ask what they can bring [to a dinner party], don't decline; instead, have them bring their specialty. Mom's chopped liver,

Aunt Shirley's carrot cake . . . they're all on the table as part of the feast. Everyone admires the fantastic food and enjoys the dinner because they've contributed to it. And they say, 'You're a perfect hostess—so relaxed!' "

Clare offers another cultural tip, if one considers "culture" in the ethnic sense. To adapt as a new member in "a warm, extended Jewish family," she advises: "First observe, and then *kiss!* Everybody—great aunts, second cousins, friends of friends. . . . Saying good-bye at family gatherings, kiss everyone twice—it's great fun, and you're a success!"

Well, listen, if it works for Clare, it may work for you. I know from firsthand experience that (a) the technique is not restricted to extended Jewish families and (b) it can sometimes require more intestinal fortitude than may be available under the circumstances. A suggested variation is the almost-official greeting at RGs and that has become so popular that it is now commonplace wherever and whenever Mensans gather: the hug. Some RGs even hold "hugging competitions," in which secret judges present awards in categories they invent themselves. I have, on more than one occasion, been the recipient of such awards, and while I am grateful for them, it's a little discouraging that they all tend to range from the avuncular to the ursine. (My size makes me a natural for great bear hugs.) The sexy awards always go to other people.

Nevertheless, hugging is a well-proven icebreaker and, when it comes to "friends of friends," considerably easier to perpetrate than kissing. It's also more sanitary.

A couple of years ago, the damnable Rubik's Cube became an ingrained part of American culture, to the delight of the Ideal Toy Company and to the consternation of Marvin A. Silbermintz, of Kew Gardens Hills, New York, who works for Ideal as a game designer and was charged with the task of coming up with saleable variations of Erno Rubik's little torture device. (You may have gathered by now that I have neither the patience nor the skill to "solve" the cube. I am convinced that the several people who gave me versions of it as birthday gifts had nothing but pure malice in their hearts.)

Ideal received thousands of unsolicited suggestions for Rubik variations, but, as Marvin says, "most of these ideas merely changed the material or superficial design of the cube."

Marvin's first response to this challenge was Rubik's Race, an ingenious combination of the basic elements of the cube and those "15" puzzles that consist of fifteen sliding tiles in sixteen spaces; the object is to arrange the tiles in some sort of sequence. (For the record, I hate those, too—and for the same reasons.) Marvin's description of how he developed the Race is too long to include here; besides, Ideal might not appreciate publicizing such things, which are, I suppose, properly regarded as industrial secrets. Suffice it to say, therefore, that he laid out the procedure for me, and it is a classical example of logical thinking in a commercial-cum-cultural context.

Marvin then solved the problem of the Rubik's Cube Perpetual Calendar by drawing on one of his cultural attributes: He was an artist before he became a game designer. His familiarity with graphics no doubt led him to such solutions as designing certain letters so that they could be used in the names of months either right-side-up or upside-down. (According to the advertising copy, "it takes only a few minutes—maybe" to come up with the right day and date every morning. I guess it's sort of a daily fix in the office for Rubik's addicts.)

Among my all-time favorite problem solvers is Harry Brown, of San Antonio, Texas, whom I described in a magazine article as "your ordinary, garden-variety genius with an uncommon appetite for problems and an uncanny ability to solve them." Harry, a true Renaissance man, became aware of a serious problem worrying the culture mavens of the world, not to mention the Italian tourist authorities. When Bonanno Pisano began construction in 1174 on what was supposed to have been a bell tower, he expected it to stand up straight. It didn't. It leaned. Furthermore, over the centuries, it continued to lean. In the mid-1970s, Italian engineers concluded that if something were not done soon, the Leaning Tower of Pisa would become the Horizontal Tower of Pisa. Here's the story, in part, as I wrote it for the magazine*:

"A call went out for the engineer . . . who could develop a practical method for keeping the familiar

---

*Cavalier, February 1976; used with permission.

phallic symbol from collapsing like a spent lover. Harry Brown read about the contest and decided to have a go at it. He went to the library, which is how he begins solving all his problems, read a few books and came up with a solution so simple that only a genius could have devised it.

"'The tower is sinking more on the south than on the north,' he explains, 'giving the effect of rotating on an imaginary axis. The problem is to stop or reverse the rotation. . . . What they have to do is pour a circular foundation around the tower, a concrete disc a hundred feet in diameter. . . . Increasing the area . . . would reduce the pressure to about 4,100 pounds per square foot. The disc should be off center. . . . The resulting differential in pressure would reverse, although very minutely, the direction of rotation.'

"So Harry drew up his blueprints and he and his wife went off to Pisa. There he received a shock. His calculations and solutions were based on the premise that the tower was set on a concrete foundation, but according to local legend, the foundation consisted of rubble. A search through Bonanno's original drawings confirmed the legend. It occurred to Harry that with that kind of foundation, the damned thing should have toppled centuries ago, but he had to set that aside to cope with the newer and more immediate problem. Installing the concrete disc would require digging away the earth around the tower's base. A foundation made of rubble would never withstand the digging. . . ."

Harry bought himself some drafting equipment,

locked himself in his hotel room for five days, and came up with a set of drawings for a plan that resembles—but with more complexity and grandeur—the method described earlier for extracting a palm tree from a hole in the sidewalk.

"It turned out, however, that Harry's original calculations had been correct. Over the centuries, underground water and seepage from rainfall had changed the rubble to concrete. He had been right all along.

"He had been right about something else, too: In Rome, he had told the Minister of Public Works that the government's fears were—well, without foundation. The tower might lean a little more, but according to his calculations, it was good for another couple of centuries at least. That prediction was [later] confirmed by an official scientific pronouncement . . . the Leaning Tower of Pisa would remain diagonally erect for another 200 to 500 years. . . ."

It should be remembered that Harry, whose checkered career runs from devising a space capsule for the first monkey to achieve orbit to designing, making, and installing stained-glass windows (his work, along with his iron sculptures, has been exhibited in museums), is self-taught. His chief attributes, in addition to his IQ, are an insatiable curiosity and the inability to resist a challenge.

Perhaps Harry Brown will inspire you as you attack the next quiz. And however you come out, remember that when you're finished, there's a symbolic Mensa hug waiting for you. Even though it may be a bear hug.

Match Wits with Mensa

CULTURE TEST

Time started _____
Time elapsed _____

GREEK AND ROMAN CULTURE

1. There is a famous river in Greek mythology which the dead had to cross by ferry. The ferryman was Charon. What was the name of the river?
2. Who was the Roman god of wine and drinking?
3. Name any of the nine Muses; take double credit if you can give the Muse's specialty.
4. Who was the individual who admired his reflection in a fountain so much that he jumped in to try to reach that beautiful person, and drowned?
5. *The Odyssey* is one of the most famous works of literature ever written. What is its main theme?

ART

1. There is a world-famous painting showing a young woman on a shell apparently rising from the sea. What is the name of the painting, for one credit? You get another point if you can name the painter, and one more if you can give his full name.
2. Most people have heard of Leonardo's famous painting called the *Mona Lisa.* What else is it called?
3. Who painted the picture of George Washington that appears on American dollar bills and on some postage stamps?
4. Which French artist is best known for his paintings and sculptures of ballet subjects?
5. There is a famous art museum in New York City that created a stir when it was built because of its unusual architecture—a spiral ramp as the main display gallery. What's the name of the building?

LITERATURE

1. Chaucer is best known for a collection of stories about the people and the life of his times. What is the book?
2. Rudyard Kipling's most famous books were chiefly about one country. Which one?
3. Name any two winners of Pulitzer Prizes for fiction since 1960. Take an extra point for each one

if you can name the book that won the prize.

4. What is generally considered to be Marcel Proust's most important work, in seven volumes? (English or French title will do.)

5. *Winnie the Pooh* has become famous in this country based largely on the later development of characters from the original books. Who was the author?

ETIQUETTE

1. How do you write a letter of regret to the White House declining an invitation to dinner?

2. Among your grandmother's possessions, you find a small tray, about 3½ by 8 inches, with a slightly raised edge. It's obviously not a candy dish. What is it?

3. What is the traditional twenty-fifth wedding anniversary title (and gift)?

4. What is the official, or approved, way of addressing a former U.S. president?

5. At a formal dinner, who is served first?

MUSIC

1. How many complete symphonies did Beethoven compose?

2. What does *a cappella* mean in relation to music?

3. Georg Friedrich Handel was born in Germany but

became famous in another country. What was his adopted country?

4. What is generally considered Wagner's most important work (or series of works)?

5. Gilbert and Sullivan wrote many operettas together. Who wrote the music?

Epilogue: Turn of the Table

Most of this chapter was drawn from several of Marvin Grosswirth's columns, "Views & Reviews" (considered a highly imaginative title when it was first begun over eleven years ago), in the Mensa Bulletin. The revival of Elstob's Foibles, as explained below, eventually led to another column, devoted to gaffes among the top 2 percent and overseen by Orville Brandes, a lady with a gravel-crusher voice, a machetelike wit, and a marshmallow heart. While working on this book, we learned, in January 1983, of Orville's death in a New Orleans hospital. This epilogue is dedicated to her, in memory of a beloved and gifted friend. Mensa, you'll recall, is Latin for "table." Orville's column was called, appropriately, "Turn of the Table."

"Boy, for someone who's supposed to be smart . . . "

Which of us has not heard, with dread, that opening of a sentence that, when concluded, revealed to all within earshot that we have managed to perpetrate an act of grandiose stupidity? We have heard it from parents, teachers, employers, lovers, spouses, and, perhaps most cruelly, from children. Anyone who has reached the age of seventeen who claims never to have heard it either is lying or simply isn't supposed to be smart. Basically, it is a case of intelligence running ahead of itself, of literally "outsmarting" itself.

There's no need to dwell on analysis. Having called your attention throughout this book to examples of how Mensans use their intelligence to solve problems, it seems only fitting that we now have a look at what happens when Mensans "oversolve."

It also seems only fitting that I begin with a tale about a garment bag. I had returned from a trip, unpacked, and proceeded to restore the garment bag to its proper resting place—the top shelf of a clothes closet. I rolled it up, tossed it on high, and watched in disbelief as it slid down again. I caught it in midair, rolled it again, and once more lofted it onto its shelf. The damnable bag sprang back at me like a thing alive, with what can only be described as naked hostility. Two or three more attempts seemed only to strengthen the bag's resolve not to stay on the shelf and to cause me physical damage at the same time.

Viciously, I grabbed the rebellious sack. I rolled it yet another time, but now I clutched it in a viselike stranglehold with one hand as I reached for a wire coat hanger with the other. Once more I flung the bag onto its shelf, and before it had a chance to unfurl, I stabbed it with the coat hanger, pinning it against the wall. My intention was to hold the bag on the shelf with the hanger while, with my other hand, I sort of stuffed the bag into position. As I did so, however, my trousers, unaccustomed to the lack of tension around my equator that resulted from so much arm-stretching, slid gently to the floor.

For a second or two, I simply froze, my arms on high and my pants bunched ignominiously around my ankles. (Mercifully, my wife was already asleep and witnessed none of this.) Finally, I looked down at my pants. In so doing, I discovered that behind the shoe rack on the floor of the closet was a perfectly good space for the garment bag. I stepped out of my trousers, kicked them aside, and held out my arms to catch the villain. I swept aside the tangle of wire coat hangers that had fallen during the fray, and I dumped the garment bag in the space, where it now permanently resides. I hung up my pants and went to watch television.

At such moments, I gain some solace from remembering a British Mensan named Peter Elstob who, some years ago, regaled readers of the *International* (Mensa) *Journal* with a tale about a padlock. Mr. Elstob had purchased what he described as "an expensive little padlock with a beautiful brass key." After the lock had served the purpose (unnamed) for which it had been bought, he decided that the most efficient thing he could do would be to store both lock and key so that he could find them easily, should the need arise to do so. "Carefully threading the U of the lock through the hole in the end of the key," he wrote, "[I] snapped it shut in order, of course, for them to remain together. They still are."

That little tale gave rise to similar stories, which soon became known, in honor of the inspirer, as Elstob's Foibles. One writer told how, before he began painting a floor, he had made certain that he had an escape route. He planned the job so that he would finish at a door in the room. It never occurred to him to check whether the door was unlocked. You know the rest. Another told how he had prepared frozen daiquiris, only to discover that some of the "finely granulated ice" would not melt. A little later, he also discovered that a shot glass was missing and developed new respect for his blender.

One particularly creative Mensan needed to make holes in a mailbox to mount a name plate. He got out his rifle and shot the holes through the mailbox. He also shot up the name plate, which he had put into the box for safekeeping.

When I related the sad, frustrating tale of the garment bag in my *Bulletin* column, I revived Elstob's Foibles and invited readers to contribute. The response was extremely gratifying.

One correspondent found it necessary to move a television antenna from one end of his one-story house to the other. Upon moving the cable, along with the stepladder that got him on and off the roof, he discovered that the line was too short. So, of course, he immediately bought additional cable. He carefully described (but I will spare you the details) the difficult and tedious task of splicing coaxial cable, involving stripping, unbraiding, soldering, insulating, re-braiding, all the while trying to keep the blood from punctured fingers from interfering with his work.

"Since I'm an engineer," he explains, "I did all that quicker than you could write three novels. So ... I grab antenna and climb to roof only to find the co-ax cable is threaded through the stepladder treads. How can I restore my wife's faith in Mensa?"

There were others: There was the woman who left her kitchen and headed for the back porch, with a cup of cocoa in each hand and cookie in her mouth. Her sweater got caught on a splinter in the door frame. With both hands occupied, she was unable to extricate herself. With her mouth likewise occupied, she could not call for help. She stood there, saying "Mfff" as loudly as she could, hoping someone would come. Someone did, freeing her while asking why she hadn't either chewed or dropped the cookie. "Mfff," she replied. There was the navy photographer's mate who left a photo darkroom carrying a stack of carefully exposed—but undeveloped—photographic printing paper under his arm, completely ruining the latent images waiting for chemical baths. There was the woman who astonished a friend who had come home with her by plopping her car keys into a teacup while simultaneously pouring boiling water into her open handbag.

Of all the foibles, my favorite is the Tale of the Ant Killer, reported in a local-group newsletter. The hero of the tale discovered ants in the kitchen and devised an ingenious plan for summarily dispatching them:

He swooped them up with a vacuum cleaner. Quickly realizing that the ants were probably still alive inside the machine, he decided to asphyxiate them by holding the end of the vacuum cleaner hose over an open, unlit gas jet. Vacuum cleaners have electrical motors. Electrical motors give off sparks. Sparks from electrical motors and gas from kitchen stoves tend to be incompatible. The vacuum cleaner blew up. Fortunately, no one was hurt, except for the ants, who are presumed dead (but may only be missing in action).

The fact that Mensans are capable of such foibles should in no way detract from the realization that we do, on occasion, come up with solutions to problems that seem insoluble to others. For example, I once protested the fact that there seemed to be an inordinate number of car alarms going off early in the morning. I am referring to those horrible sirens that are supposed to scare off anyone tampering with the hardware of a parked automobile. Maybe they accomplish their goal, but they also destroy sleep and engender a sense of irritation that persists through the day. My solution for doing away with this problem was to do away with the time during which it occurred.

I was less tolerant in those days than I am now, so I pointed out that the only people abroad between the hours of 4:30 and 5:30 A.M. were hunters, fishermen, people waiting for the golf courses to open, muggers working overtime, and car thieves, and I characterized them all as undesirable. I then urged everyone to write to his or her government representative, demanding that the hour in question be abolished. "When those noxious sixty minutes cease to exist, the undesirables whose time it is will be unable to function," I said.

But Stan Bercovitch, a long-time friend and now head of the Mensa local group in Fort Myers, Florida, said: "I have always felt that the morning was much too close to the night. By abolishing the hour between 4:30 and 5:30, you are bringing the two even closer together."

He was quite right. The solution, however, is beautiful in its simplicity. Rather than being discarded, those minutes should be inserted between 6:00 and 7:00 A.M., making a 120-minute hour. Everything else is being recycled; why not time? Just think what it would mean to have 120 minutes to spend between six and seven o'clock in the morning. The longer you think about it, the more plausible the idea will become, especially when you are getting ready to go to work.

Another solution to a long-standing problem was published in several local newsletters and is offered here with the permission of its author, John H. David, of St. Louis. It is included here not to draw ridicule; on the contrary, it is such an elegant piece of logical thinking that it is an eminently appropriate note on which to conclude this book.

Jack David proposes the abolition of presidential elections, on the premise that in an ailing economy, with both the Republicans and Democrats making claims and counterclaims, we don't know whom to believe. With a presidential election looming, matters are bound to get worse, with, as he puts it, "the added burden of media coverage of every nose twitch of every possible candidate.

"I think I have a solution for both of these situations, although there are a few bugs left to be worked out in the system," he admits.

"Reduced to its essentials, the system posits that we would dispense with the presidential election and put the government out for bids. The winning party can decide for itself who should be president.

"Of course we would have bidding specifications, such as: (1) a top limit for the national debt, (2) a guarantee that unemployment would not go beyond a certain percentage, (3) adequate defense against invasions by foreign powers, (4) the maintenance of the Social Security system at a specified level.

"I imagine we would have thirty or forty such items.

"Congress would meet just once every four years to set up the specifications. The results should be reasonable, since each side would want to be able to meet them.

"Control of the government would go to the lowest bidder. Since our tax structure would be based on the low bid, this should work out well for the taxpayer. Think of all the money we could save in the electoral process alone, not to mention all the other dollars that the present procedure costs.

"And, perhaps best of all, we could require a performance bond, something we've never had before."

Sounds good to me, Jack. Now all we have to do is convince Congress.

When you write to your congressperson in support of Jack's proposal, why not mention the 120-minute hour, too?

ANSWERS

TRIVIA WARM-UP ANSWERS

1. The word *piccadil* was applied to the slashed collar worn by dandies who strolled in the area that is now Piccadilly. A house called Pickadilly Hall, built by a tailor named Higgins who made a fortune selling these collars, was built on the street, according to a 1656 citation.
2. Percy Bysshe Shelley was the husband of Mary Wollstonecraft Shelley, who wrote *Frankenstein*. (The monster himself is never named, incidentally; Frankenstein is his creator.)
3. Sandhurst.
4. Wine grapes.
5. A baronetess. She is addressed as "Lady"—for example, Lady Jones. (A baroness is the wife of a baron.)
6. A woman, under the applicable British law, takes her husband's rank. Queen Elizabeth II holds her title in her own name. A husband does not take his wife's rank under the same circumstances.
7. A morganatic marriage, in which the wife and children do not take the husband's rank or right to succession.
8. Switzerland. It's the abbreviation for Confederation Helvetia. Helvetia is the country's Latin name.
9. O'Hare Airport, Chicago, Illinois.
10. John Harvard and Harvard University.
11. English. There had been English fishing boats

along the New England coast for many years, and fairly good evidence indicates he had been to England, taken there by one of the fishing boats.

12. From *planta genesta,* the broom plant. Geoffrey, Count of Anjou, adopted the use of a sprig of this plant as a symbol during a pilgrimage. His descendants used it as a family name and ruled England from 1154 to 1485.

13. The color titian, a brownish orange, was named after Titian (1477–1576).

14. Verona, Italy. Although the lovers of Shakespeare's play did not really exist, their "homes" are on the standard tourist route.

15. Pocahontas.

16. Arabia. (Don't confuse him with the writer D. H. Lawrence of *Lady Chatterley's Lover* fame.)

17. Saturday's child must work for a living.

18. In Mr. McGregor's garden.

19. The Ninth Symphony.

20. Queen Elizabeth I.

21. *"To Anacreon in Heaven."*

22. In Dublin's fair city.

23. Chile.

24. The Boer War.

25. In Agra, India.

TRIVIA TEST ANSWERS

The figure following each answer is the percentage of Mensa members who got the correct answer. Nobody got them all right. The average number of right answers was fourteen; the average time taken to complete the test was nine minutes. The highest score was nineteen correct.

1. In the pancreas. (Credit is also given to the British member who said, "Near St. Pancras Station," which is a railway station in London.) (57%)
2. Mata Hari. (More than 14% thought it was Gertrude Stein, who would probably have been amused. "Name is a name is a name.") (11%)
3. Rhodesia. (68%)
4. A novel that uses real people, thinly disguised, under other names. The French term means "novel with a key." (25%)
5. One example is *Cakes and Ale* by Somerset Maugham, in which, it is believed, Thomas Hardy and Hugh Walpole were satirized. There have been a good many novels in the last few years that appeared to be fictionalized, thinly disguised lives of personages such as movie or stage stars and Greek shipping tycoons. (21%)
6. (a) a shoe. (55%)
7. A garden in Kashmir. (7%)
8. Janus, the two-faced Roman god of gates and doorways. (89%)
9. Fishing. (61%)
10. Printing. (82%)
11. Peanut. It is also sometimes called a monkey nut. (71%)
12. Northeast, to the Atlantic. (79%)

13. They are all spas, locations where hot or mineral springs exist and are used. (93%)
14. *Life with Father.* (36%)
15. Probably from the Clink Prison, on Clink Street in Southwark, London. The street still exists, and there is a sign indicating the former site of the prison. (There is some question whether the street was named for the prison or vice versa; another possible derivation is from *clinch*, an old English or Dutch word meaning "lock up" or "fasten securely.") (4%)
16. (b) (91%)
17. Perennial. An annual may reseed, but perennials are supposed to come up every year. (79%)
18. From Nimes in France. The cloth was originally known as "de Nimes" or "from Nimes." (58%)
19. Athos, Porthos, and Aramis, and D'Artagnan (no credit lost for spelling the names incorrectly). (39%)
20. Village smithy. Not so incidentally, if your answer was *blacksmith*, you were not correct. Smithy is the building, blacksmith is the person, and the poem says very clearly that it was the building standing under the tree. (50%)
21. In the Southern Hemisphere sky. (Credit is also given for answering that it is found on certain flags.) (86%)
22. Mauritius. (21%)
23. Louis Daguerre, of daguerreotype fame. The moral of this question may be that if you are going into partnership with anyone, make sure *you*

have the more pronounceable name. (11%)
24. March 15. (Several people also included 44 B.C., but there were Ides every March. It should also be noted that the Ides occur on the 15th only in March, May, July, and October. The other months of the year have their Ides on the 13th.) (71%)
25. The Trylon and Perisphere. (32%)

VOCABULARY WARM-UP ANSWERS

1. (c) The dinner was a *medley* of recipes from around the world.
2. (d) The guest who insulted his host was *injudicious*, to say the least.
3. (e) A *deputation* from a group of workers in the plant visited the management office.
4. (d) Garlic is reputedly an *antidote* for werewolves.
5. (d) She volunteered for the dangerous mission entirely of her own *volition*.
6. (b) There are many sports that seem to be *hazardous* to the participants.
7. (c) When I asked for a logical explanation, all I got was a *rigmarole*.
8. (a) The rules of the boarding school tend to *circumscribe* the conduct of the students.
9. (b) *Empirical* judgment is often sound, but a little study helps, too.

10. (d) To refuse to shake hands with someone is a direct *affront.*
11. (a) The child who was refused permission to play outside became *petulant.*
12. (c) The courtier was *assiduous* in his attentions to the new king.
13. (a) The actress's *sardonic* smile showed her disbelief of the story told to her.
14. (b) When the usually shy, sedate young lady became a go-go dancer, her behavior was totally *incongruous.*
15. (b) Many science fiction stories and novels relate to *chronal* discrepancies.
16. (c) Religious meetings are often opened with an *invocation.*
17. (b) The butterfly was particularly colorful, with *iridescent* gleams on its wings.
18. Don't look a gift horse in the mouth.
19. A stitch in time saves nine.
20. Too many cooks spoil the broth.

VOCABULARY TEST ANSWERS

The average number of correct answers was fifteen. The highest correct score was nineteen. (That person missed *obiter.*) The average time was six minutes. The figure in parentheses following each answer is the percentage of Mensa test takers who got it right (rounded off).

1. (c) The miser had an inordinate amount of *cupidity*. (75%)
2. (b) The *enormity* of the mass murder shocked the nation. (There is almost unanimous agreement among standard authorities that *enormity* does *not* mean "very large" and should not be used in that sense.) (40%)
3. (c) *Progenitor* can be used literally or figuratively, to refer either to a person who is the source of an idea or to an actual ancestor. (100%)
4. (c) You may *careen* a ship by turning it over, or a wagon may *careen* from side to side, figuratively showing the underside; but a car that is speeding out of control is *careering*. (33%)
5. (c) The *nefarious* criminal was stopped in his tracks by the hero, who was wearing a white hat, of course. (92%)
6. (d) Many of the scenes in television situation comedies are intended to be *risible*. (92%)
7. (d) Many herbal products and chemical irritants have been used to produce *aperient* products. (25%)
8. (d) The *diaconate* of the church was occupied alternately by a man and a woman. (50%)
9. (c) An *obiter* (or *obiter dictum*) is an opinion expressed in passing and, if given by a judge, has no effect on a ruling. (33%)
10. (a) A restaurant named Sam's Place uses the owner's name as an *eponym*. (75%)
11. (d) Executed pirates, as described in old stories, are usually said to be hanged from a *gibbet*. (50%)

12. (b) Anything occurring during the period of the equinoxes, which is when the days and nights are of equal length, can be described as *equinoctial.* (75%)
13. (e) Many bodily functions are considered to be *autonomic.* (60%)
14. (b) A specialist in a field is generally *erudite* in that particular subject. (100%)
15. (b) A *dour* person is usually not pleasant company. (100%)
16. (b) The forger was *mendacious,* as well as criminal, in his actions. (56%)
17. People who live in glass houses shouldn't throw stones. (100%)
18. Early to bed and early to rise, makes a man healthy, wealthy, and wise. (96%)
19. A fool and his money are soon parted. (96%)
20. Don't change horses in the middle of a stream. (75%)

LOGIC, REASONING, AND MATHEMATICS WARM-UP ANSWERS

1. 7-22-13. This is a simple substitution cryptogram in which you substitute letters for numbers: 8 = S, 22 = E, 7 = T, etc.
2. ALL THAT GLITTERS GETS STOLEN.

3. Twenty-two unicorns with 1 horn each; eleven rams with 2 horns each.
4. (d)
5. B
6. Lettuce, celery, and roast chicken.
7. 9 A.M.
8. Avers; Saver.
9. *Kindness* or its equivalent; the word is CRUELTY.
10. 25 gallons.
11. 20¢ (10¢ per vowel).
12. (b)
13. 16
14. D 31 (December 31; the series started with July 31).
15. II
16. Lying—because no liar would admit to being a liar.
17. 20 hours.
18. Y
19. Small change.
20. One-way (one-weigh).

LOGIC, REASONING, AND MATHEMATICS TEST ANSWERS

The figure in parentheses following each answer represents the percentage (rounded off) of Mensa members who got that question right. The highest number correct was nineteen, and that individual finished in ten minutes. The average number right was almost fifteen (a small fraction under), and the average time was thirteen minutes, with the exception of problem 12. Many people took more than an hour to work that one out, if they did not know how to do it initially. An explanation of how to solve puzzles of this type is included with the answer, as requested by many people.

1. (a) All of the others may or may not be true, but you cannot tell from the information given. The oldest grandchild could be a boy, for example. (65%)
2. THE EARLY WORM GETS WHAT IT DESERVES—EATEN. (50%)
3. Five—one more than the number of colors. (80%)
4. 9 A.M. (84%)
5. (e) FIG (alphabetical order). (88%)
6. Coat, $100; skirt, $50; scarf, $10. (96%)
7. Liars; rails. (36%)
8. (c) (92%)
9. 6 (84%)
10. A (100%)
11. VEIN. It's a simple substitution cryptogram. (44%)
12. 1089 × 9 = 9801. (24%)
   The puzzle and explanation were provided by Warren Sears, of Beachwood, Ohio.

$$\begin{array}{cc} 1089 & \text{ABCD} \\ \underline{\times 9} & \underline{\times 9} \\ 9801 & \text{DCBA} \end{array}$$

It is reasoned out as follows: 9 × A must equal a one-digit number. So A can only be a 1 or a 0. Trial and error prove that A = 1; 9 × D must equal a number ending in 1. The same trial and error demonstrate that D = 9; 9 × B must equal a one-digit number, but 1 has already been used. Therefore, B = 0. Next, 9 × C (+ 8 carried from 9 × 9) must equal a number ending in 0. By the same trial-and-error method, it is possible to demonstrate that C = 8.

13. Bravery (scrambled word is COWARDICE). (24%) (Take credit for *courage* also.)

14. 15. It is the only one of the numbers shown that is not a prime. (48%)

15. COPENHAGEN. It is the only city not on the North American continent. (44%)

16. Alice and Tom, Sally and Bob, Gertrude and Bill, Betty and Jim (or, of course, the same lineup in reverse order). (44%)

17. D (60%)

18. 208 (32%)

19. 66⅔% (52%)

20. (d) (56%)

ANALOGIES WARM-UP ANSWERS

1. Red Sea and Mediterranean Sea.
2. (d) *a*. It is the only letter that is above the line, as *p*, *q*, and *g* are below the line.
3. (a) discrete.
4. (b) story novel.
5. Chicago.
6. Automobiles.
7. Rosinante (the animal to which the folk hero is linked).
8. A
9. President John Quincy Adams (father and son).
10. Underwater exploration.
11. Chopin.
12. George Orwell (pen name).
13. Paris—both are what we call the subway. In England, a subway is an underground walkway below a street, usually used for getting across a dangerous intersection.
14. South Dakota (state capital).
15. Football.
16. D
17. Bishop.
18. Kansas City.
19. Great Britain.
20. Gabriel D. Fahrenheit.
21. The Philippines.
22. Ghana.

23. Russia.
24. Hockey.
25. *My Fair Lady* (Oscar winners).

ANALOGIES TEST ANSWERS

The average number of correct answers was fourteen. The highest score was eighteen. The average time was ten minutes. The figure in parentheses following each answer is the percentage of Mensa members who achieved a correct score on that question.

1. The United States. Liederkranz cheese was developed by Emil Frey in the 1890s and named for his singing group (*Liederkranz* is German for "wreath of song"). It is a U.S. trade name. (4%)
2. Fear of light. Photophobia is often due to illness or unusual sensitivity. (80%)
3. Potatoes. Parmentier introduced potatoes to French cuisine, so if you don't like potatoes, don't order anything à la Parmentier. (8%)
4. Newark, New Jersey. Those are the initials used for the airports. (64%)
5. Anna (the word reversed). (64%)
6. Pod (or small school or herd, but *pod* is the technical word). (40%)
7. *Pygmalion* (the source of the musical comedy). (36%)
8. Lewis Carroll (pen name). (36%)

9. Africa. (92%)
10. Heterodox (the exact opposite; but *unorthodox* is also an acceptable answer). (44%)
11. (a) strengthen. (68%)
12. Morocco. (4%)
13. D (96%)
14. (a) fraudulent. (92%)
15. (d) revealed. (84%)
16. C (52%)
17. Isolde or Iseult (or any variant of the same name). (92%)
18. (a) Harding. (64%)
19. (a) Venice. (So called because condemned prisoners crossed it after sentencing.) (56%)
20. (c) sand. (72%)
21. (c) Truman. The first Roosevelt (Theodore) succeeded McKinley and became president on McKinley's death, as Harry S Truman did after the second Roosevelt's (Franklin Delano's) death. Note: There is no period after the S in Truman's name because he had no middle name, and the S is not an abbreviation. (16%)
22. (b) Sri Lanka. (68%)
23. (d) fountain. (16%)
24. (b) cooking. (8%)
25. (a) 1, 11, 14. This is comparable to warm-up analogy question 2. Here, the first set of numerals (3, 5, 8, 9) are considered as curved lines, so the only set that would match 4 would be the numerals written with straight lines. The warm-up question used letters similarly, but it is the same general idea. (12%)

CULTURE WARM-UP ANSWERS

ART

1. The chapel was named after Pope Sixtus IV and was built as a papal private chapel in 1473. There are many other famous paintings in it besides those of Michelangelo.
2. The pictures depict the life of the upper-class courtesans who were an important part of Japanese male social life of the time.
3. Sculpture.
4. Raphael was also an architect. He designed, among other buildings, the Palazzo Pandolfi in Florence, Italy.
5. Gibbons was a sculptor and woodcarver who did a tremendous amount of work at St. Paul's Cathedral and many other places.

MUSIC

1. D'Oylcy Carte presented the operettas at the Savoy Theatre in London.
2. *La Marseillaise.* Claude Joseph Rouget de Lisle composed the words and music, but it was originally written as a war song for the Army of the Rhine. On well-documented evidence, on July 30, 1792, a troop of volunteers from Marseilles entered Paris singing the song, which was thereafter called by its current name.

3. Egypt.
4. No, because the waltz wasn't invented until almost two hundred years later.
5. New Orleans.

GREEK AND ROMAN CULTURE

1. Zeus and Hera.
2. Romulus and Remus.
3. The Wooden (or Trojan) Horse.
4. Lesbos.
5. Spinning thread. Clotho held the distaff, Lachesis spun the thread (of life), and Atropos cut off the thread when life was finished.

LITERATURE

1. They were played by boys.
2. Eddas (the Poetic Edda and the Prose Edda).
3. Blindness.
4. *The Divine Comedy.*
5. The Arthurian legends—the life and death or disappearance of King Arthur.

ETIQUETTE

1. The flute-shaped glass, since the bubbles dissipate more slowly. Of course, the champagne "saucer" is also correct.
2. The gentleman is always presented to the lady, unless the gentleman is the president of the United

States, the recognized head of another country, royalty, or a high church dignitary.

3. Never. According to etiquette books which cover the subject, a woman does not take off her gloves to shake hands, and does not apologize for them.
4. Nothing. It is considered more courteous to write the message, even if it is only a few words.
5. According to most authorities, at home at the family dinner table, or at a picnic.

CULTURE TEST ANSWERS

The average number of correct answers by Mensa members was just under eighteen out of a possible thirty. (Some questions have several parts.) The average time was twelve minutes, with a low of eight minutes and a high of thirty-nine minutes. The figure in parentheses following each answer is the percentage (rounded off) of Mensa members who got that answer correct. There was one question, number 4, on Greek and Roman Culture, that everybody got right.

GREEK AND ROMAN CULTURE

1. River Styx. (80%)
2. Bacchus. (85%)
3. Calliope, Muse of epic poetry; Clio, Muse of history; Erato, Muse of love poetry; Euterpe, Muse

of lyric poetry; Melpomene, Muse of tragedy; Polyhymnia, Muse of sacred poetry; Terpsichore, Muse of choral song and dance; Thalia, Muse of comedy and idyllic poetry; Urania, Muse of astronomy. (50% named one or more; 45% were able to give the Muse's specialty. Among Mensans the best known Muse was Terpsichore.

4. Narcissus. (100%)
5. The adventures of Odysseus (Ulysses) on his long and epic journey home from Troy. (60%) (Two other answers given were *Greece on $5 a Day*, and *You CAN Go Home Again.*)

ART

1. *The Birth of Venus* (the usually accepted title) by Sandro Botticelli (a.k.a. Alessandro di Mariano Filipepi). (60% knew the name of the painting; 45% knew the painter; 5% knew the painter's first name.)
2. *La Gioconda* (Italian) or *La Belle Joconde* (French). (40%)
3. Gilbert Stuart. (35%)
4. Edgar Degas. (60%)
5. The (Solomon R.) Guggenheim Museum, designed by Frank Lloyd Wright. (70%)

LITERATURE

1. *Canterbury Tales.* (90%)
2. India. (90%)
3. 1960, *Advise and Consent* by Allen Drury; 1961,

To Kill a Mockingbird by Harper Lee; 1962, *The Edge of Sadness* by Edwin O'Connor; 1963, *The Reivers* by William Faulkner; 1964, none; 1965, *The Keepers of the House* by Shirley Ann Grau; 1966, *Collected Short Stories* by Katherine Anne Porter; 1967, *The Fixer* by Bernard Malamud; 1968, *The Confessions of Nat Turner* by William Styron; 1969, *House Made of Dawn* by M. Scott Momaday; 1970, *Collected Stories* by Jean Stafford; 1971, none; 1972, *Angel of Repose* by Wallace Stegner; 1973, *The Optimist's Daughter* by Eudora Welty; 1974, none; 1975, *The Killer Angels* by Michael Shaara; 1976, *Humboldt's Gift* by Saul Bellow; 1977, none; 1978, *Elbow Room* by James Allan McPherson; 1979, *The Stories of John Cheever* by John Cheever; 1980, *The Executioner's Song* by Norman Mailer; 1981, *A Confederacy of Dunces* by John Kennedy Toole; 1982, *Rabbit Is Rich* by John Updike. (And, of course, any added to the list since then.) (70% could name two authors, 20% could name the book of one author, 5% could name both authors and both books.)

4. *Remembrance of Things Past,* or *À la Recherche du temps perdu.* (35%)

5. A. A Milne. (65%)

ETIQUETTE

1. You don't—it's a command performance except for an absolutely major reason, such as your own wedding or funeral. (35%)

2. It is most probably a visiting-card tray, for calling cards. (50%)
3. Silver. (95%)
4. Officially, just plain "Mr." Usually, out of courtesy, "Mr. President," when you speak to him. (He is spoken of as "the former president.") (50%)
5. Woman guest of honor, to the host's right. The host used to be served first, according to some texts, to prove to his guests that the food was not tainted or poisoned. At present, service is as given, except for certain circumstances, such as a single hostess. (50%)

MUSIC

1. Nine. (95%)
2. Voice alone, without accompaniment (literally, "in chapel or church style"). (55%)
3. England, where he became a citizen. (50%)
4. *The Ring Cycle,* or *The Ring of the Nibelungs,* or *Der Ring des Nibelungen.* (75%)
5. Sir Arthur S. Sullivan. (30%) (He also wrote "The Lost Chord," a famous hymn.)

About the Authors

ABBIE F. SALNY, Ed. D., is a psychologist licensed in the State of New Jersey, and a diplomate of the American Board of Professional Psychology. In private practice, she specializes in supervising doctoral trainees (i.e., people who have received their doctorates but need practical experience) and in diagnostic testing.

She is a retired professor and deputy chairman of the psychology department at Montclair (NJ) State College. A Mensa member since 1964, she serves as Supervisory Psychologist for both American Mensa and International Mensa. Her husband, Jerome E. Salny, a retired business executive, is also a Mensa member. That's where they met.

MARVIN GROSSWIRTH is a journalist and author whose books and articles cover a wide variety of subjects, with a propensity for science, technology, and medicine. He joined Mensa in 1961 and has served as the organization's public relations officer continuously for twelve years, except for a two-year term as national chairman. He is married to Marilyn S. Grosswirth, a technical writer, whom he met at a Mensa party. They both expect their son, Adam, to join Mensa soon.

How to Join Mensa

If you would like to join Mensa, you can find out how by writing to Mensa, Dept. AW, 2626 E. 14th St., Brooklyn, NY 11235. In Canada, write to Mensa, Dept. AW2, P.O. Box 505, Station S, Toronto, Ontario M5M 4L8. You'll need a score on an IQ test that's at the 98th percentile or higher. Mensa can help you take such a test. If you've already been tested, they'll tell you how to submit "prior evidence."